From Outpost to Outport

From Outpost to Outport

A Structural Analysis of the Jersey-Gaspé Cod Fishery, 1767–1886

ROSEMARY E. OMMER

McGill-Queen's University Press
Montreal & Kingston • London • Buffalo

© McGill-Queen's University Press 1991
ISBN 0-7735-0730-2
Legal deposit first quarter 1991
Bibliothèque nationale du Québec

Printed in Canada on acid-free paper

This book has been published with the help of a
grant from the Social Science Federation of Canada,
using funds provided by the Social Sciences and
Humanities Research Council of Canada.

Canadian Cataloguing in Publication Data

Ommer, Rosemary E.
  From outpost to outport
  Includes bibliographical references.
  ISBN 0-7735-0730-2
  1. Fish trade – Quebec (Province) – Gaspé – History.
  2. Fish trade – Channel Islands – Jersey – History. 3.
  Cod-fisheries – Economic aspects – Quebec (Province)
  – Gaspé. 4. Gaspé (Quebec) – Commerce – Channel
  Islands – Jersey – History. 5. Jersey (Channel Islands)
  – Commerce – Gaspé (Quebec) – History. I. Title.
  HD9464.C33Q8 1991   338.3'72753   C90-090318-X

This book was typeset in 10/12 Palatino by
Typo Litho composition inc., Quebec.

For Andy, Catriona, Ken, and Keith

Contents

Tables

Figures

Acknowledgments

The text for this book came into existence in 1979 as a Ph.D. dissertation for the Department of Geography at McGill University under the supervision of Sherry Olson, whose contribution is hereby gratefully acknowledged. Since then it has undergone many revisions, under the guidance of anonymous readers and with the help of the late David Alexander and Keith Matthews, of Memorial University of Newfoundland, who gave unstintingly of their time and wisdom. I want to thank my friend, Patricia A. Thornton, of Concordia University, for her constant encouragement, for advice on matters demographic, and for help with mapping and the tabulation of data. Gregory S. Kealey and Alan G. Macpherson, both of Memorial University, read an earlier draft of the manuscript and offered editorial advice; along with Peter Goheen of Queen's University, they provided needed support and encouragement during the preparation of the book. I am particularly grateful to Daniel Vickers, of Memorial University, whose detailed editorial and contextual comments and criticism at several stages in the writing process were invaluable. I must have tried his patience sorely, but he never let it show.

My thanks and appreciation are due to many people in Canada and Great Britain. In Canada, I wish to thank Larry McCann (Mount Allison University) and Gordon Handcock and William Schrank (Memorial University of Newfoundland) for useful criticism, advice, and information; and Ernie Forbes (University of New Brunswick) for his encouragement. I owe a special debt to the late Mr and Mrs Arthur LeGros of Paspébiac for the interviews they granted me, and for permission to study the documents of Charles Robin and Company (CRC) in their private possession. Mr LeGros was an expert on the firm, and he generously shared with me many of his insights, adding enormously to my knowledge of the firm and the trade. I also wish to thank Gloria de Blazio and Dorothy Savoie, both of Montreal, for interviews they granted. I am likewise grateful to David McDougall and the late Mrs McDougall, of Montreal, as well as Clarence LeBreton, of the Acadian Village, Caraquet, for the many fruitful discussions I had with them on the subject of Gaspé itself as well as on CRC.

In Jersey, my thanks go first and foremost to the staff of La Société Jersiaise in St Helier. In particular, I was helped enormously by Barbara deVeulle, Margaret Syvret, Joan Stevens, Henry Perrée, Helen McCready, and John Appleby. I also wish to thank Mr Antil and Mr Falle, of the Jersey Library, the Honourable Secretary of the Jersey Chamber of Commerce, and Mr Palmer, of the Bureau des Impôts, Jersey, and his staff. Philip deVeulle, Frank LeMaistre, Jennifer Mullins, Reg Nicolle, John Norman, Mr Gates, and Mr Slauenwhite (the last two of the Royal Trust of Canada, Jersey Branch), Lewis Gibaut, Mary Clement-Robson, Mr and Mrs Guy Janvrin Robin, Nicholas Robin, and Lady Angela Walker all granted me interviews and treated me with courtesy and hospitality. This book would, quite simply, not have been possible without the assistance and interest of all of these people. My thanks are also due to the staffs of Custom House, Quality House, and the Public Record Office, all in London, England. Likewise, I received expert assistance from Ms Holmes of the Dorchester Record Office, and the staffs of the Mitchell Library (Glasgow), the Maritime History Archive (Newfoundland), the Public Archives of Nova Scotia, and the National Archives of Canada.

On a more personal level, I wish to thank Mr and Mrs Ashborn (St Helier), John and Maura Mannion (St John's), the late Mr E.P.G. Thornton and Mrs Thornton (then of London), Andrea and Vincent Brogan (Glasgow), Ted and Margaret Palmer (Poole), Pat Thornton and David Shortall (Montreal and Ottawa), Bill Smith (Ottawa), and Jim and Milesza Gilmour (Ottawa) for their hospitality and assistance "in the field." Jim Gilmour must bear the responsibility for having introduced me to export-base theory, although what I have done with it since then is not his fault. I am grateful to Chris Palmer for taking excellent care of four small children while I did fieldwork. My thanks go, as well, to Robin Craig, of St Margaret's Bay, for hospitality, encouragement, and many valuable lessons on ships and shipping.

I am grateful to Routledge, Chapman and Hall, Inc., *Business History*, the Maritime Studies Research Unit of Memorial University, and *Acadiensis* for permission to reprint material previously published by them. The research for this work was funded by a McConnell Fellowship and two summer fellowships from McGill University, by a Bourse du Québec from the Gouvernement du Québec, and by a travel grant from the Institute of Social and Economic Research at Memorial University of Newfoundland. Thanks are due to all these bodies, as well as to Michael Staveley (Dean of Arts at Memorial) for financial assistance with cartography, and the Social Science Federation of Canada for the funding which enabled this book to be published. My heartfelt thanks go to Beverly Maher, who typed the final version of the manuscript, and to Colleen Dalton, Irene Whitfield, Joan Butler, and Diane Dawson, who, over several summers carefully and patiently typed earlier drafts. They kept me going with their interest and good humour. Cliff Wood, Gary McManus, and Dennis Curtis of the Geography Department and the Cartographic Laboratory at Memorial (MUNCL) designed and drew the maps and carefully refrained from expressing the horror they must have felt when they saw my original sketches; I am grateful to them for their professional skill and their personal tact.

My thanks are due also to Phillip Cercone, executive director of McGill-Queen's University Press, Joan McGilvray, co-ordinating editor, and Käthe Roth, who copy edited the book, for making this an enjoyable and satisfying publishing experience.

Finally, I want to thank Anne Alexander, Andrea Brogan, Mike Browne, Moyra Buchan, Phil Buckner, Gale Burford, Marjory Campbell, Judy Fingard, Sandi Galloway, Gerry Hallowell, Colin Howell, Kathleen Hugessen, Martine Johnson, Linda Kealey, Zita Kelly, Kay Matthews, Lewis MacLeod, Stuart Pierson, Mickie Pitcher, Frank Remiggi, and Mark Shrimpton for being there in their own particular ways when I needed them. This book is dedicated to my children – Andy, Catriona, Ken, and Keith – who, sometimes gracefully, sometimes reluctantly, but always with love, tolerated the intrusions on their time that a project like this involves. They may not always have my time, but they have my unconditional love.

From Outpost to Outport

Introduction

This is a book about fish: the people who caught it, the people who bought and sold it, and the places in which this activity took place in the late eighteenth and the nineteenth centuries. In more scholarly terms, it is a study of a colonial staple commodity trade: how it was established, how it worked, why it worked the way it did, and how all of that affected the economic development of the mother country and its colonial hinterland. In 1766, Gaspé became an outpost of the Jersey metropole. In 1886, Jersey turned its back on Gaspé, and the region was reduced to Canadian outport status. This book charts the history of the economic relationship that existed between those two places as bounded by those two events and seeks to explain it.

It is now fifty years since Harold Innis wrote his monumental and seminal study of the Canadian Atlantic fisheries, which he identified as belonging to an international economic trade.[1] Since then, surprisingly little historical work has been done on the structure of that trade, although the modern east-coast fisheries have attracted more attention, especially in recent years.[2] Yet the roots of the current major issues in the inshore fishery, and of some of the problems of eastern Canadian regional underdevelopment, lie in the way in which the early merchant fishery was first established and conducted; at least, that is one main contention of this book. Consequently, these pages do not contain a detailed chronological account of the history of the merchant inshore fishery at Gaspé,[3] nor do they offer a comparative study with other Atlantic Canadian fisheries (inshore or banks). Instead, the purpose of this work is to provide a structural and theoretical analysis of a trade which expressed in economic terms the relationship between a metropole and its colony: the emphasis is on structure and process rather than on event.

Jacob Price has pointed out that, of the four ways to go about studying colonial economic history – "the empire, the colony, the community and the commodity" – none is satisfactory.[4] The empire approach underestimates the impact of the market system; the colony approach tends to neglect the "external aspects of the economy"; the community approach is even more isolationist in practice; and the commodity approach, while the most promising, is just "too difficult," since the canvas upon which the history of a commodity (for instance, wheat, iron, cotton, fish) has to be painted is so large. Too much is involved, from the statistics of the trade itself to the populations that consumed the commodity and the international political and economic arenas within which the trade had to operate.

The task is indeed daunting – well-nigh impossible – if treated as Price proposes. There is, however, a way out of the dilemma: the use of the case study as a way of simultaneously cutting the canvas down to manageable proportions while preserving the essential detail and complexity that alone can aid our understanding of the specifics of a trade. So this book looks at the fish trade using one firm, one metropole, and one colonial region as a surrogate for the whole. It is not so much a perfect as a proximate solution, of course. Some detail gets lost in the process (such as comparative information on other regions), but the relative order of things is maintained and the relationship of one part of the commodity trade system to another is revealed, even at the international level.

The particular firm used as case study in this book is Charles Robin and Company (CRC), of Jersey and Paspébiac. There are four reasons for this choice. First, present knowledge of the Atlantic fisheries is almost entirely based on studies of Newfoundland and the West Country connection, while the history of the Gulf fisheries and the Jersey firms remains relatively unexplored.[5] Second, the Jersey firms were dominant in the Gulf of St Lawrence fisheries in the nineteenth century.[6] Third, CRC was the pre-eminent Jersey fishing firm in the New World fisheries. Fourth, the records of this firm are more complete than are those of any other Jersey firm. As a major Jersey firm, CRC was genuinely representative of the nineteenth-century Gulf of St. Lawrence inshore cod fisheries.

There is a tradition among scholars of referring to an international trade that has three landfalls as a "triangular trade": the trade that took manufactured goods from the United Kingdom to Africa, slaves from Africa to the West Indies, and sugar back to the United Kingdom is perhaps the best-known example. The fish trade was also triangular – a triangle of function rather than of landfall. There were three discrete spatial foci of activity, each with its own particular function. One was the home (management) base – the metropole – in this case the island of Jersey. The second was the production base, broadly defined as the Gulf of St Lawrence littoral and more specifically as Gaspé and the outports. The third was the marketing arena, best thought of as a far-flung network or range of options, including the Mediterranean, the West Indies, and Brazil, that were geographically diverse but functionally identical. Throughout this book, this functional merchant triangle is used as a device for conceptualizing the Jersey cod trade, because it accurately reflects the manner in which management, production and processing, and marketing were actually carried out by the merchants themselves. Because the three apexes of the trade were geographically distinct, merchants had to deal with them as separate foci of effort, linked by money and information flows, by lines of control within an individual firm, and by the spatial movement of goods and services. They were also managed by merchant-derived policies and strategies designed to organize a staple trade in such a way as to gain access to a trading cycle of exchange of goods which ultimately returned to the Jersey metropole in the form of cash. The information structure required to operate this merchant triangle was vast, almost globe-circling, but it was ordered from control points and it generated triangular voyages.

Using this merchant triangle as an analytical device, the various functions of the different sectors of the trade can be identified, the connections between them recognized, and the balance between them kept in mind; at the same time the problems of each function can be examined separately, and the implications of solutions sought at any one apex for any other part of the system can be dealt with coherently. Beyond this, the difficulties of managing a commodity trade across considerable distances and intervals of time can be captured and the inherently international nature of such trades be truly appreciated.[7]

Finally, in a study which must range over a large number of areas of research each one of which represents major challenges in its own right, the idea of this "merchant triangle" allows the reader to keep track of temporal and spatial changes as they occurred and to relate them to one another, without becoming lost in a web of detail. For example, the Jersey fish trade in Gaspé was in existence from 1767 to 1886, and its antecedents evolved on both sides of the Atlantic over a two-hundred-year period at least. The canvas on which the story of this trade is painted must be large enough to include the origins of the trade, the move to Gaspé, the dominant years of the firm, and its demise. The wider context is also required, including the role of Jersey in the British scheme of the fish trade, the international markets for the trade, the value of the trade to Jersey, and its impact on Gaspé. Only then can the logic of the evolution of this commodity trade become comprehensible.

That said, three discrete elements have been identified as key to this study: how and why the Jersey fish trade developed as and when it did; the difficulty of organizing such a trade profitably given the common-property nature of the staple; and the economic and managerial interplay between metropole and colony and its impact on development in a newly settled region. These are the major themes of this book.

Chapter 1 deals with the metropolitan rise of the cod fishery in Jersey from the time when that island had a small rural economy based on domestic agriculture. Chapter 2 shifts to the production apex of the trade with the establishment of Charles Robin and Company on the Gaspé coast. Solutions to the problem of setting up and maintaining a fish business on this coast in the late eighteenth century are examined here. Pre-eminent among the difficulties Robin faced was the problem of dealing with a mobile common-property resource over which he needed to establish proprietorial rights if he was to create a viable enterprise. His response was to design a series of measures that gave him control over the resource through the acquisition of control over the fishermen. The consequent forms of labour control are discussed at length in this chapter against the more general backdrop of indenture and indebtedness. Other problems are considered as well, such as the setting up of markets and the difficulty of operating in an international arena in wartime.

With the trade established, chapter 3 returns to the metropole to consider how Jersey, as a periphery to Great Britain's fish trade, could create a secure niche for itself within the larger metropole. Skillful manipulation of constitutional ambiguities and the institutionalization of merchant solidarity in the creation of the Chamber of Commerce were vital components in Jersey's success at the metropolitan end of trade. Chapter 4 then looks at the trade in action, to see how it functioned as an established and successful business during the central period of Jersey hegemony in the Gulf of St. Lawrence.

Chapter 5 returns to Gaspé to consider the impact of the established trade on the producing colony. In essence, the problem posed here is to discover how the system initiated by Charles Robin in the late eighteenth century affected the economic development of the region. Central to the answer is an analysis grounded in export-base theory as developed by Baldwin, North, Watkins, and Gilmour.[8] The normal linkage effects of the fish staple are shown to have been thwarted by the mercantile strategies which secured metropolitan success, and in the process laid the basis for regional underdevelopment.

Chapter 6 therefore poses the reciprocal question: How did the fish trade affect the economic development of the metropole? This is a logical extension of the old staples approach,[9] but one which, perhaps surprisingly, has not been systematically pursued by students of export-base theory. When the question is applied to nineteenth-century Jersey, it becomes clear that the vital backward and final-demand linkages of the fishery were captured by Jersey and used to foster ancillary industry, shipbuilding, and, eventually, a world-wide carrying trade.

No study of this kind is complete without some indication of what ultimately happened, and chapter 7 provides this in what is essentially an epilogue. The merchant triangle is shown to have come under increasing stress after mid-century, and the whole system finally collapsed in 1886. Chapter 8 then brings the various themes and issues of the book together in an essay on colonial staple trades and regional development.

The essential question posed by this book is: How did the cod fishery, functioning as a commodity trade, shape the economic development of the metropole that managed it and the colony that produced it? The literature relevant to such a necessarily sharp focus is surprisingly slight, especially when the staple model is recast as a commodity-trade model. While Price's comment that "there is nothing new about this [staples] approach, at least since Innis' day" is unfair to North, Baldwin, Watkins, and Gilmour, it is nonetheless true that, these four apart, "serious scholars have done relatively little with [staples]."[10]

This book, literature on staples aside, has tended to work within the framework of a generalized development approach, or ethic, rather than to borrow from any particular school of development thinking. I have a general, rather than a specific, debt to Tom Naylor, Daniel Drache, André Gunder Frank, David Alexander, and others who have cared so deeply about regional and colonial underdevelopment.[11] My debt to those who have laid the empirical foundation of Atlantic Canadian fisheries studies – Innis, Matthews, Ryan, and Head,[12] in particular – is likewise of a general nature, more accurately acknowledged in this way than in the form of a review of what they have achieved. Together, the work of scholars such as these provides the moving spirit behind this book.

The answer to the question of the impact of the fish trade on regional development has been formulated here within the context of a "special case" only insofar as the case-study approach has been adopted to make the problem manageable. The implications of this analysis, applied to other trades and other regions, may well be considerable. Certainly, with respect to eastern Canada, this study has relevance even in the last decade of the twentieth century in the context of a perceived failure of the eastern Canadian inshore fisheries to provide an adequate livelihood for local fishermen. In Jersey, the fisheries of the Gulf of St Lawrence set that island on a path to economic development based on the careful nourishing of captured linkages and the consequent creation of a viable "invisibles" sector of the domestic economy. Today, Jersey enjoys a lucrative finance industry based yet again on other countries' wealth and transnational locational decision making. In the process, as classic export-base theory would predict, the old exogenous staple base in Gaspé was sloughed off. In Atlantic Canada, in contrast, the heritage of the merchant cod fishery was the staple trap, rather than economic development. Today we are faced with the paradox of enduring regional poverty in the face of what was, until very recently, resource plenty. It is the roots of this dilemma that are explored in this book.

PART ONE

Origin and Structure of the Trade

CHAPTER ONE

Beginnings

In 1837, a writer in the *Guernsey and Jersey Magazine*, observing the thrust of Jersey colonial expansion into the British North American cod fisheries, noted that "the industry of a nation rests not so much on the extent of its territory as on that of its capital,"[1] an encouraging point of view for people living on an island that was only forty-five square miles in area and not particularly rich in natural resources. Indeed, given the diminutive proportions of its "national" territory, Jersey could never realistically have hoped to develop great wealth or to support an expanding population merely from the fruits of domestic economic development. For expansion of the island economy to occur, outside support of some description was needed. It is therefore not surprising that Jerseymen seem to have been traders from very early in their history. Nor is it surprising, since trade flourishes under conditions of political stability, that they were also consummate political strategists. They had to be, for Jersey, although a possession of the English Crown, lies in the English Channel just off the coast of France, and from the Middle Ages into the nineteenth century England and France were often at war.

The geographical ambivalence of the Channel Islands is not only reflected in their French name – Les Îles Anglo-normandes – but also in their constitutional position within the United Kingdom. In 931 AD they were annexed, along with the Cotentin (on the French mainland), to the possessions of the French Duchy of Normandy. When the seventh duke, William, was crowned King of England in 1066, the Islands became connected to England as part of the estates of the English monarchy. When King John inherited them he instituted royal courts in both Guernsey and Jersey, thereby laying the basis of the present Jersey system of central administration based on the feudal ideal.[2] From Norman times onward, all land in Jersey belonged to the king in his capacity as Duke of Normandy, and all estates were granted to families in return for services to the Crown. These "seigneurs" in turn granted portions of land to tenants in return for services rendered to the lord of the manor.

Also from the time of King John, Jersey was gaining experience in fish as a trade good. There was a flourishing trade in salted conger eels which were sold to Europe in large qualities: the records of the Crown duty (*esperkeria*) levied on these eels and mackerel show that the Channel Islands exported eight thousand congers annually during King John's reign. Such a reliance on foreign trade might appear to have been dangerous, given the virtually incessant political and economic conflict between France and England, but Jerseymen became adept at protecting their island from wartime depredations while deriving economic advantage from its geographical location on the interface of hostilities. During the Hundred Years' War, for example, the commerce of Jersey actually prospered. Visits from the English Navy provided a steady market for local wheat, fish, and butter, while foreign markets were not damaged, since Jerseymen were able to extract safe conducts for their trading vessels from both French and English admirals as a result of the strategic value of the island to both fleets. Indeed, in 1483 Edward IV of England and Louis XI of France agreed to regard the Channel Islands as neutral territory in any war between their two countries, and in 1484 that agreement was ratified by Pope Sixtus IV. Even after the Reformation, Elizabeth of England kept this "privilege of neutrality" among those she bestowed on Jersey, declaring that "in time of war, merchants of all nations, aliens as well as natives, friends and foes, can without impediment, frequent the Islands to escape storms or for purpose of Commerce and Admiralty, without molestations, and remain in safety so long as the Island remains in sight."[3]

During Elizabeth's reign, Jersey flourished as a royally encouraged entrepôt for Anglo-French trade, which could not otherwise have been conducted in safety, since Elizabeth's support of the French Huguenots had brought her to the brink of war with France. Also during Elizabeth's reign, the trade in knitted goods (from which has come down to us the famous "Jersey" sweater) became an important part of the island's commerce, reportedly gaining popularity as a result of the queen's personal predilection for Jersey hose. A local coasting trade with British ports developed, based on the importation of such goods as canvas and coal and the exportation of Jersey cloth. By the time of the reign of James I, the knitting industry had become so important that the Royal Court of Jersey was forced to decree that "during harvest and the Vraicing [gathering seaweed for fertilizer] season, all persons should stop making stockings and work on the land on pain of imprisonment on bread and water and the confiscation of their work."[4]

This order is a clue to the dilemma that the islanders faced if demographic and economic expansion was to be feasible. By the early 1600s, the population was about 25,000 souls and the old purely agricultural economy was being replaced by a more diversified economy that included fishing, growing apples for cider production, and knitting as well as agriculture. These economic activities were carried on within the general context of a feudal system, with a "free" peasantry whose obligations to the seigneur were discharged mostly by payment in kind.[5] The seigneurs held between them about 100 to 130 fiefs, were addressed by the names of these fiefs, and were referred to as "noblesse." Below them in social status were the peasant farmer-proprietors producing wheat, dairy produce, vegetables, apples, and fish. Until the sixteenth century, wheat was the main crop, and surplus was exported to Normandy. In Jersey, economic power was based on land, but land was extremely limited. By 1624, only eighteen years after the Royal Court's prohibition on knitting during periods when the land needed attention, commerce was on the rise and the English parliament was being petitioned for larger supplies of wool because the knitting trade had become the sole means of support for more than one thousand persons.[6] At the same time, land was enclosed and used for apple orchards, not, according to the local historian deGruchy, by fiat of the seigneurs but by general consensus. Enclosure, deGruchy argues, was seen by a free peasantry as being more economical, since the profits to be gained from exporting cider more than made up for the loss of common pasturage. By 1682, Jersey had ceased to export corn and now imported "at least half of its needs,"[7] and deGruchy summarized the process thus: "The main basis of the Feudal System, the collaboration of groups of husbandmen with a lord to produce food and clothing for their own subsistence, had gone, and farming for profit had taken its place."[8] Nevertheless, he pointed to an attempt at retrenchment on the part of some seigneurs as the old system declined.

There had been other developments as well over the previous hundred years. During the reign of Elizabeth, Jersey became directly involved in the cod fisheries of North America. In the early sixteenth century these fisheries were being exploited by Norman and Breton boats, and oral tradition maintains that Jerseymen frequently crewed on vessels sailing out of Saint Malo and other neighbouring French ports. The earliest reference to a Jersey-based involvement in the cod trade is a will of 1582 in which Pierre de la Rocque left his sons equal shares in the ship "which is now unloading after she voyaged to Newfoundland." In 1587, the Royal Court dealt with a dispute over a cargo of Newfoundland cod, and in 1591 one Jean Guillaume was fined three hundred crowns because he sold his Newfoundland fish at Saint Malo.[9] This Jersey-Newfoundland involvement continued, and probably expanded, in the seventeenth century, although the degree of participation is masked by the fact that many Jerseymen and Jersey vessels of that era were listed as being out of places such as Weymouth, Lyme Regis, Dartmouth, or Southampton.[10] Interestingly, these fisheries, along with other small local ones, operated outside of any feudal restraint, since "no general feudal privilege over fisheries seem to have existed."[11]

At the political level, too, Jersey was experiencing a widening of horizons – a point that can be illustrated through the history of the deCarteret family of the mid-seventeenth century. The deCarterets were an old Norman family whose original estates had been on the Cotentin and who had amassed a considerable amount of political power in Jersey by the time of the English Civil War. Philippe deCarteret, seigneur of St Ouen's, was both bailiff (having chief responsibility for the civil and judicial administration of the island) and lieutenant governor (responsible for defence and the monarch's representative on the island); his brother Elie was solicitor general (serving under the bailiff); his cousin Helier was attorney general (serving under the solicitor general); three other cousins and one nephew were jurats (who exercised judicial and legal functions under the Royal Court).[12] Of the twelve captains who commanded the militia, seven were deCarterets, two more were nephews, and one was a brother-in-law. Philippe had the king's promise that when he died his brother Elie would be the next bailiff, and his nephew George would succeed Elie in the post. Not surprisingly, the de-Carteret family was instrumental in keeping Protestant Jersey on the side of the Royalist forces during the Civil War. George de-Carteret supplied them with munitions from France and captured supply ships heading for London (a Parliamentary stronghold), re-selling their cargoes in Saint Malo in order to buy Royalist supplies. Philippe was also involved in this venture; in 1643, when Parliament ordered his arrest, George took over the running of the island, engaging in privateering until he acquired a small fleet with which to serve the king. These freelance privateers were known as "pica-roons," and Cromwell retaliated against one of their raids by capturing ten Jersey fishing vessels that were at the Newfoundland fishery. In 1649, Charles II took refuge on Jersey, which subsequently came to be known as a pirate's (that is, Royalist) lair. On the island, the monarch had always been credited with the power to heal scrofula by the laying on of hands, so when Charles came to Jersey, twelve sick men were brought to him to be healed.[13] Their subsequent cure is said to have confirmed the islanders' belief in the legitimacy of Charles' claim to the throne and to have guaranteed their absolute loyalty thereafter. Jersey caution before investment in an "adventure" apparently extended beyond the purely commercial!

In 1660, with Restoration, came reward both for the deCarterets and for the island. George was given several positions at Court and became one of the eight proprietors of Carolina. He also received what is today known as New Jersey and was one of the six persons to whom the King granted the Bahamas. Jersey, through the deCarterets, thus acquired at least a formal transatlantic focus, with a "proprietary colony" on American land granted by the king in return for military services rendered by the seigneurs of the principal fief of the "realm" of Jersey.

Under Charles' royal munificence, and with help from income derived from privateering, trade expanded and the island grew prosperous. At this time, St Aubin (later to become the principal port of the cod trade) developed: its pier was built and a market was instituted to be held every Monday "for merchants of foreign commodities."[14] By the 1660s agriculture had lost its dominance, and cider making, knitting, and fishing were on the increase. In 1689, smuggling, which had become something of a national sport in Jersey, prompted William of Orange (now King of England) to withdraw the ancient privilege of neutrality, although locally his prohibition was regarded as nothing more than a tiresome temporary interference with trade. Balleine observed that smuggling into England was almost as profitable as smuggling into France, and noted that the lieutenant governor himself was "almost always in bed. He commands a public commerce with France by vessels which arrive and depart at night."[15] During the late seventeenth and early eighteenth centuries, with the expansion of the volume of shipping moving between Britain and her empire, Jerseymen (like inhabitants of many of the West Country towns) were able to employ shipping in a range of transatlantic trades.[16] The fishery continued to develop slowly, and not until 1731 was it referred to as the prime trade of the island. In that year seventeen vessels sailed for Newfoundland with 1,500 seamen, and twenty-seven vessels sailed the following year. By 1771, "45 ships were annually employed in the Newfoundland fishery," according to Shebbeare.[17]

The relatively slow rise of merchant capital in the Jersey–New World fisheries requires some explanation. Part of the reason undoubtedly lies in the economic implications of the Jersey system of power-holding. Power in Jersey was rooted in the land; there was no monopoly control of any kind over the cod fisheries. Moreover, seigneurial interest in transatlantic commerce (as opposed to land rents) seems to have been negligible: New Jersey, for example, was sold at the death of courtier George deCarteret by his executors, to clear his debts, without having stimulated any noticeable interest in transatlantic commerce. Rather, rent returns from colonial possessions seem to have been the main concern of the deCarterets,[18] and it appears that Jersey seigneurs, involved in securing financial returns from their lands in Jersey or abroad, were not motivated to venture into the risky and unknown world of transatlantic enterprise. In other words, population growth in Jersey seems to have led to the export of labour in the form of seamen who wanted to make their sea-going enterprises ultimately part of the Jersey economy – that is, Jersey-based. The early transatlantic fisheries may therefore be thought of as clinging to the edges of the domestic land-based economy, while the later fisheries were a result of the growth of these minor ventures into major enterprises which became central to the island's economy. Other transatlantic episodes, then, like those of deCarteret, were just that: episodes, side issues, not part of the economic evolution of the island.

The cod fisheries of Newfoundland (see figure 1) were undoubtedly among the riskiest of transatlantic "adventures," since they were chronically insecure from their inception, at the very beginning of the sixteenth century, until the middle of the eighteenth century.[19] English and French involvement in Newfoundland was always based on the cod fisheries that had been prosecuted off the shores of that island by a number of European nations since Cabot's rediscovery of it in 1497. By 1507, fish were being procured in the area by English, French, and Portuguese vessels, with Spain following closely thereafter. In the seventeenth century, with the Spanish and Portuguese fisheries in decline, the major competition was between the French and the English, with the English fishing along what is now called the "Old English Shore" and the French concentrating on the south and west coasts and the area north of Bonavista Bay.

There were some basic differences between the traditional French and English fisheries at this time. The French evolved a "pêche errante," which consisted of catching fish and salting them down on board the fishing vessel – the "green" cure – thereafter returning with the catch to the mother country. The English generally used a "dry" cure, which involved salting and sun-drying on shore during the season. The former technology required no attachment to the land, while the latter did: one was a banks fishery, the other inshore.

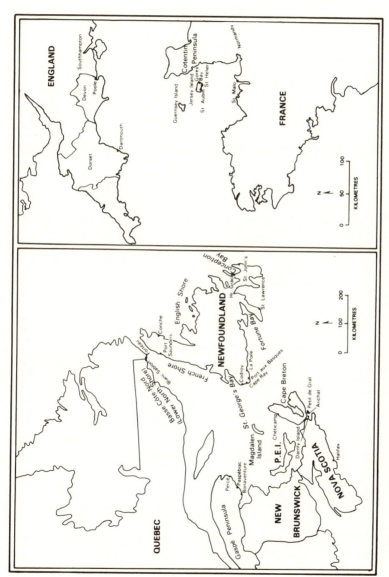

Figure 1. Jersey and the British North American Fisheries

This meant, in effect, that the English fishery from its inception carried with it the prerequisites for settlement, although for a variety of reasons it took a long time for permanent settlement to actually become established.[20] The French fishery, on the other hand, evolved no significant settlement until after the Franco-Spanish War (1657–59), when France began to devote a great deal of attention to the fisheries. An even more fundamental distinction between the English and continental fisheries was that there was no large sustained market in England for dried Newfoundland codfish.[21] The English fishery was therefore inherently triangular in its trade patterns, since marketing had to take place in areas outside Great Britain.

In 1661, having claimed sovereignty in Newfoundland for a long time, the French finally put teeth in that claim with the establishment of a colony at Placentia. The French bank fishery increased steadily, as did the dry fishery on the French Shore.[22] A French protective tariff barrier raised in 1678 made trade into France almost impossible for English goods; England retaliated with similar tariffs. From 1689 to 1698, and again from 1701 to 1713, French and English fisheries, settlers, garrisons, and navies were in conflict in Newfoundland. By 1713, the fortunes of war on land and sea had gone against the French and, in the Treaty of Utrecht, France finally recognized English sovereignty over the entire island, while retaining the right to catch and cure fish between Pointe Riche and Cape Bonavista during the fishing season.

During the second half of the seventeenth century, for a complex of reasons, the English fishery at Newfoundland fell into a steep decline.[23] Some metropolitan ports formerly prominent in the trade disappeared completely. Others barely survived, including Jersey. In 1670, Jerseymen were to be found only in small numbers in Trinity Bay, complaining that French competition was restricting their trade since the French were able to "navigate their ships at a cheaper rate," and objecting to "the levying of a duty of a crown per quintal in France" upon English fish brought in there. As a result, "the number of Jersey vessels employed in that trade declined from twenty to three or four."[24] Falle, writing in 1694, bemoaned the drastic effect of French privateering on the Jersey fleet, saying that matters had come to such a pass that "the neighbourhood of St. Malo, that famous retreat of French corsaires, has received our navigation."[25] After 1713, however, the situation appears to have improved slightly. With the south coast ceded to the English, their fishery could be extended geographically. By the 1730s, Jerseymen were solidly established in Conception Bay, the DeQuettevilles having become accepted as "an old time Newfoundland family" with an establishment on the south side of Harbour Grace called the Jersey Room; by 1775, for example, Nicholas Fiott, a Jerseyman, had become the fishing admiral at Harbour Grace.[26] Indeed, the fortunes of the English fishery as a whole tended to improve steadily from the 1730s onward, although its greatest expansion did not occur until after 1763, with the end of the Seven Years' War.[27] The Treaty of Paris made the whole coast of mainland British North America accessible to the English fishery, and French-speaking Jerseymen, who were well used to visiting fishing sites in the Gulf of St Lawrence as crew on St Malouin fishing vessels, were peculiarly able to take over these places. Very soon after the Treaty was signed, Jersey firms were to be found in Gaspé (Robin, Pipon and Company; Janvrin and Company; Fiott and Company) as well as in Newfoundland, where, "of the 34 sail of shipping in Conception Bay in 1763, one third were from Jersey, the Custom Port with by far the largest representation."[28]

The Jersey fish trade, then, grew along with a general expansion in the English fishery at Newfoundland; but, in importance and numbers, Jersey remained basically peripheral to the main English fishing centres which were based in the West of England. The expansion of Jersey fishing enterprises into the Gulf of St. Lawrence and along the coasts of Newfoundland (see figure 1) tended to distance Jersey merchants from the West of England fishing centres on the old English Shore of Newfoundland; it also seems to have prompted these merchants to start organizing themselves politically and commercially in Jersey itself.

T he need for such organization arose out of the arrangement of political and economic power in Jersey, which in turn resulted from the peculiar constitutional relationship of the island to the United Kingdom. Jersey was a possession of the English crown in the monarch's capacity as Duke of Normandy; it was not clear what the island's position was with respect to the British parliament at Westminster. Indeed, the problem was still exciting debate as recently as 1977, and Jersey's constitutional position remains "undefined and vulnerable to changes of policy by ... British Governments."[29] The real question has centred around the location of the seat of legislative power ever since the constitutional monarchy became a reality in Britain. Since Jersey's links to England arose through its obligations to the monarch as Duke of Normandy (that is, the King in Council) and not to Westminster (that is, the King in Parliament), the latter's power to legislate for Jersey was debatable. Jersey was not simply a colony or dependency of England, and it could be argued that it was not subject to restrictive colonial legislation. In the eighteenth and early nineteenth centuries, particularly with a mercantilist British parliament enacting legislation to protect its trade, and with a constitutional monarchy established in Britain, the Jersey parliament (the States – les États) found itself in a position where the British parliament could argue that the island was not a part of Britain, while the States could respond that it was "neither a colony nor a conquest, but a peculiar and immediate dependency of the Crown, and in all commercial treaties the Channel Islands have always been admitted upon a footing of equality and as one with the people of the United Kingdom."[30]

Who was it, however, that would speak for Jersey and its cod trade at Westminster? The normal instrument of contact between England and Jersey was an order of the King in Council which was ratified by the States of Jersey. As the case of the deCarteret family demonstrates, the States operated as part of a political hierarchy, which indeed it was, since it had grown out of the Royal Court, originally a judiciary body which in the course of time had acquired joint legislative powers in Jersey with the King in Council. The States were concerned, among other things, with the smooth economic functioning of the island within the general constitutional framework evolved since 1394 and reaffirmed in 1716 by George 1 (3 Geo 1, C.4). The problematic position of Jersey had been underlined by the accession of William III, who was not a hereditary Duke of Normandy but a legally constituted successor under a recognized parliamentary supremacy. George 1 confirmed Jersey in its traditional right (previously confirmed by Royal Charters) of free import into England of its produce and goods and provided a reciprocal arrangement for England. In 1771, the States commenced legislating in its own right following an Order in Council to that effect.

The political power of London was of particular importance to Jersey in any economic endeavours that islanders undertook in British colonial possessions overseas, and Jersey expansion in the cod fisheries of British North America was necessarily predicated on a favourable, or at least non-combative, attitude to such ventures in Westminster. While such was normally the case, Jersey fish merchants needed the security of a voice in the corridors of power that would be able and willing to cope with emergencies, since, as Matthews observed of the West Country merchants in the eighteenth century, "at the first hint of war, piracy, proposals to settle a governor in Newfoundland, 'bad' laws, or even the growth of Jersey or Irish trade, the merchants instinctively cried for help from the Government."[31] The States, the official politico-economic channel for dealing with such problems, was not likely to come to the assistance of the fisheries in the early years of the trade, given the seigneurs' lack of interest in transatlantic endeavours, even in the American colonies in which they held proprietorial rights. Their interests seem to have remained essentially monetary, land being seen as a means of financial gain through quitrents or sale. It was the merchant venturers following in their wake who had set up transatlantic commerce, not the seigneurial courtiers themselves. In Jersey, the interest of the States lay in winning land-based power struggles such as the reaffirmation of the right to sell the produce of its lands in the English markets. Indeed, as late as 1770, some of the landed aristocracy were attempting to increase seigneurial wheat "*rentes*" returns by artificially creating wheat shortages. This scheme resulted in rioting in Jersey to protest the Act of the States, which, by authorizing corn exportation, had created the "shortages."[32]

The fisheries, of course, did not fit this pattern of economic power based on land, since the cod fishery was a marine occupation dependent on an exogenous resource and subject to exogenous control in the form of the British Trade Commissioners, the Custom House, and the King in Council. In effect, this new venture represented a bifurcation in the economic thrust of Jersey, a new development and one in which the landed interests on the island had no expertise. Jersey merchants wishing to operate in such an enterprise therefore had to devise a second system of power and influence which would enable them to exploit their exogenous marine resource successfully, since the States was concerned only with domestic and land-based affairs.

The peripherality of Jersey merchants to the Newfoundland–West Country fishery, to the land-based power of the States of Jersey, and to Westminster was circumvented four years after the Treaty of Paris with the establishment of the Jersey Chamber of Commerce, the first such institution in the United Kingdom.[33] The purpose of the Chamber[34] was to allow Jersey merchants to speak with one voice to both the States and Westminster, by providing a vehicle that would minimize conflict and competition among themselves in order to deal with it effectively elsewhere.[35] It was essentially a problem-solving device for the merchant class, described by its creators as being designed to promote "the well-being of trade and to support and keep the merchants upon a respectable footing."[36] The members of the Chamber hoped that this forum would keep them united in policy and mutual aid, thereby making it an effective voice for the island's trading interests in the corridors of power. Such a pressure group required good politico-economic contacts, and these were quickly established. The Chamber contacted Jersey merchants resident in London, requesting "that they will take upon themselves the management of the affairs of the Chamber and that they will among themselves fix on one or more of them who may particularly receive our instructions and remittances from time to time."[37] Letters were also sent to Lord Albemarle, the governor of Jersey, and to Lord Granville, the bailiff of Jersey, seeking their protection. Finally, the Chamber wrote to agents in London and Southampton asking that they, as well as masters of Newfoundland vessels, accord freight preference to its members.[38]

Subscriptions to the Chamber were "three pence sterling per ton per annum ... to be raised upon the tonnage belonging to each respective member ... not less than thirty tons" and those who did not own vessels or vessel shares subscribed according to an estimated tonnage which they thought appropriate. The thirty-ton limit kept out owners of very small vessels (such as local coasting boats or fishing smacks) unless they had more than one vessel, which was highly improbable. In fact, the majority of the initial subscribers were Jersey merchants involved in the cod trade: thirty-two out of a total initial membership of fifty. They included Patriarche, Lemprière, DeGruchy, LeBreton, Pipon, Robin, Winter, and Janvrin – a roll call of the principal Newfoundland cod trade merchants.[39]

This first Chamber of Commerce was short-lived and collapsed in 1772, according to the Minute Book. Although the background of the collapse is not clear, it may well have been a result of the success of the expanded Jersey fisheries, which would have made merchant cooperation and solidarity appear unnecessary to some merchants, leaving them "unwillingly to join in maintaining the commerce of the Island [of Jersey] upon a proper footing,"[40] as the Minute Book reported. During the early 1770s, this was quite likely: in 1771 Jersey ships were at Conche, Codroy, Cape Ray, Fortune Bay, and St Lawrence, and by 1775 the Channel Islands were sending "not less than 60 or 70 vessels and 1550–2000 men annually" to the fisheries.[41] But in 1776 the English fishery was in trouble again, its expansion interrupted by the American Revolutionary War (1776–83). The war posed enormous problems for the whole fishery, but especially for the bye boats[42] and the migratory fishery. Merchant vessels and the bank fishery were vulnerable to sea attacks, and the bye-boat keepers were badly hit by labour impressments and captures going to and from Newfoundland.[43] With the entry of Spain into the war, in 1779, all European markets but Portugal were closed until hostilities ended in 1783.

The reopening of markets saw another rapid expansion in the English fisheries, including those of Jersey. As had happened at the beginning of the expansionary period after 1763, Jersey merchants reactivated the Chamber of Commerce to facilitate this new opening of the trade. They immediately tackled a perceived problem of French bounties, since they thought that "the effects of the same will operate in favour of the French codfishery to the great prejudice of the trade of this Island and all other ports in His Majesty's dominion from whence the Newfoundland fishery is carried on."[44] The panic was unnecessary, however, for from the Treaty of Versailles onward the French fishery steadily declined, falling from 273 vessels in 1774 to a mere forty-six in 1792, not to recover until after 1815 and the end of the Napoleonic Wars.[45] Some of the Jersey merchants were also outraged by their exclusion, as part of the Treaty, from the French Shore. Jersey fisheries at St George's Bay and Humber River were stopped in 1784–5, and at Port aux Basques and Port Saunders in 1786,[46] and it seems that such prohibitions may have been the reason for Jerseymen moving to locate on the Labrador coast, an expansion which was underway by 1790. In that year, for example, DeQuetteville, having survived a serious market glut in 1788, was fishing in LaPoile, Jersey Bay, and Blanc Sablon; by 1806 he had effectively monopolized the Jersey fishery at Forteau Bay.[47]

After 1783, then, the merchants of Jersey were once again a unified body expressing solidarity through the Chamber of Commerce. With this organization in place, the basic metropolitan framework existed from which the Jersey cod-fishing firms could begin to develop their emporium in the Gulf of St Lawrence. By 1785, Jersey was sending seventy to eighty vessels to the new Gaspé establishments.[48] Philip Robin and Company was flourishing at Arichat, and its agent, Charles Robin, was at Paspébiac intent on creating what would become the most important Jersey fishing company in the New World, an exercise in merchant capitalism which would survive for one hundred years.

CHAPTER TWO

Production and Markets: The Fishery at Gaspé

With the vacation of the Gulf of St Lawrence by the French as a result of the Treaty of Paris, its inshore fishery became a new precinct for British fishing firms, very much open to exploitation by whichever interests could find a way to develop such sparsely settled territory. In 1765, Jacques Robin, having failed to establish a farming and fishing settlement two years earlier at Miramichi,[1] was fishing at Arichat, calling at outports on nearby Darnly Island and Petit Degrat, and maintaining a summer station at Chéticamp. His enterprise – called Robin, Pipon and Company of Isle Madame – was starting to compete internationally against other well-entrenched fishing interests, especially those in Newfoundland, but to be sure of success it needed to secure large quantities of fish for export on an annual basis. This required expansion of the firm's catching area, and so the following year Charles Robin, as agent for the company, left Arichat for a preliminary foray into the Baie des Chaleurs to investigate its potential as a new base of operations in the Gulf. The Baie, however, was occupied by only a handful of planters and Indians (there were only 209 people in Gaspé in 1765[2]), and it was already the domain of a few Quebec-based merchants. Robin's problem was how to establish and maintain a viable merchant fishery in the area, given the existing competition for a small number of local fishermen.

The story of Robin's early years on the coast has been told in some detail by David Lee.[3] In 1767, Robin returned to Paspébiac to establish a headquarters there for the family company, now styled "at Isle Madame, coast of Acadia and at Paspébiac in the Baie of Chaleurs, coast of Canada."[4] Carrying with him enough wood to build a stage, he arrived in the Baie on the 2nd of June on board the *Recovery*, to find a Halifax vessel already trading in the area and a

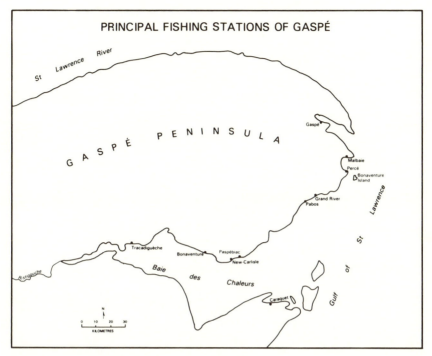

Figure 2. Principal Fishing Stations of Gaspé

government surveyor at Restigouche. Their presence was symp-
tomatic of two other difficulties he would have to overcome: competition from external traders and the development of political
infrastructures that might work for or against him, depending on
his ability to manipulate them successfully. During the following six
weeks he set about establishing contacts with local planters, trading
in Caraquet, Shippegan, Bonaventure, and Tracadiguèche as well
as Paspébiac, selling needed salt and supplies in exchange for fish
and furs, and dealing with both planters and Indians (see figure 2).
He did a satisfactory business that year and, over the next few years,
found a niche for the company in the area, winning planters and
dealing with competitors when necessary. With one, for example –
William Smith, a Quebec-based merchant – he entered into an agreement to divide the Gaspé coast between them, Robin acquiring the
area "from Paspébiac down."[5] Other early enterprises seeking to
develop the coast failed, including the Quebec firm of Moore and
Finlay and Alexander Mackin, whose operations closed in 1769, and
some "Halifax interests" which had started up at Bonaventure but
collapsed after three years.[6]

Slowly, Gaspé began to grow. By 1774 there were 158 people at Bonaventure and another 200 scattered around the Baie. In 1776, Nicholas Cox (the newly appointed lieutenant-governor of the District of Gaspé, who was subordinate to Governor Haldimand of Quebec) reported three families at "Gaspee" and four on Bonaventure Island, two at each of the seigneuries of Grand River and Pabos, and ten families of about six persons each wintering at Paspébiac and joined each spring by another thirty people, mostly from Jersey, for the prosecution of the fishery. Malbaie and Point St Peter (near Gaspé) were "inhabited by people from the Rebel Colonies who came away at the commencement of the War," and there were also Acadians at Bonaventure and Tracadiguèche – "a sober and industrious people ... improving and cultivating the land" as well as fishing. The *Census* for that year shows a total population, seasonal and permanent, of 874 persons on the coast between Gaspé and Tracadiguèche, a picture typical of early permanent settlement in a fishing area where occupancy figures were often inflated by transient single males.[7]

In 1778, the American Revolutionary War brought a serious disruption of trade to the Baie. American privateers harassed the fishing establishments and Robin was forced to abandon the coast: "The country seems ruined; operations for next winter must be laid aside, and all property removed in the fall."[8] By September he had left for Jersey, not to return until 1783.

When Robin returned, he immediately applied for a grant of land for himself and for the settlers with whom he had done business prior to the war. An influx of refugees was beginning to swell the population of the eastern Gulf – 435 Loyalist settlers came to Paspébiac in 1783[9] – and Robin was concerned that enough land be left "for the use of the fishermen."[10] He was, in fact, facing institutional opposition from Governor Haldimand and Felix O'Hara, Cox's agent on the coast, who were concerned to prevent any trade monopoly in the Baie. In 1784, Robin complained to Cox of the proposed township survey of Paspébiac that it would "cut off the fishermen at Paspébiac from the wood for flakes, stages and fires and compell them to go for a supply to Nova Scotia, or abandon the fisheries to the persons in the neighbourhood"[11] – that is, that the fishery would either have to start spending money on timber, which would not pay, or be reduced to a local subsistence activity. Robin thought that "a frontage of nine miles would be required to furnish timber for the use of the fishery"; the surveyor, Felix O'Hara, agreed somewhat reluctantly with his general concerns, writing to Haldimand that "unless in every case a proper reserve is made for the fishery it will inevitably dwindle."[12] O'Hara, however, informed Cox that, while "Robin deserves encouragement," it was important "to guard against monopoly of lands in a settlement that will become extensive."[13] Lee notes that once Haldimand left for England, in 1784, matters became easier for Robin; by 1786 he had acquired legal control of "the best part of the best beach on Chaleur Bay," in effect becoming the only merchant to operate a private beach, since shortly thereafter waterfronts became part of the public domain.[14] By this time, Robin was acting as an agent for himself and his nephews, to whose father he wrote many years later, when he wished to retire, "Do you forget how loath I was to renew the business in '83, that finally I did it for about six years, that your children might continue it?"[15] Whether his motives at the time were genuinely altruistic or not, he was in fact in the throes of creating the firm of Charles Robin and Company – CRC – whose name survives on the coasts of the Gulf of St Lawrence to this day. Then, however, such longevity could not even be dreamed of, and CRC was but one of several firms that would attempt to secure a business niche in the area, albeit a firm headed by an unusually astute and determined merchant.

Entrepreneurial vision was crucial to commercial survival in the fishery in those early years. There were a host of problems that had to be dealt with, some local, some external, some foreseeable, some utterly unpredictable. Fishermen had to be found and supplied on a regular basis, European markets penetrated, and competition from other merchants on the coast – a combination of Guernseymen, Jerseymen, and Quebecers – disposed of. Robin's letters are full of complaints about these matters as well as about other, more esoteric concerns, such as "the high duties on molasses and rum ... the difficulties inherent in Gaspé's greater distance from Europe ... the ice in the spring and fall, and ... having to compete with Newfoundland, which paid bounties to bankers (26 Geo III c.26) and admitted rum and molasses free."[16]

Competition was the most serious problem for any fish business that was not yet securely established. In 1790 Robin was worrying about external competitors, expressing his concern that "large numbers of American vessels secured foreign registries in Halifax and participated in the Chaleur Bay Fishery."[17] In a letter of 1792 to his London agents, Fiott deGruchy, he spelled out the problems of internal competition on the coast. This letter pointed to the importance of land ownership in the fishery in a newly settled area, which, in the early days, was the crux of the problem: "Messrs Matthew Stewart and Company who have carried on considerable business on this coast for some years are meditating to purchase next winter

... the seigneurie, a post that furnishes us yearly at least 2000 quintals of fish ... If they succeed we shall have a rent to pay of at least 26 quintals of fish ... But the principal consequence will be that the Seigneur will have all the fish of the post."[18] Nor were Matthew Stewart and Company Robin's only local rivals. In 1794, "a Guernsey Employ was set up in Gaspee Bay, under the agency of Thomas LeMesurier; soon after another was set up at Percee by Nicholas Fiott and Co. under the agency of Geo. LeGeyt. Another on the Island of Bonaventure by Hamond, Dumaresq and Co. under the agency of Peter LeBlanc; one at Malbaye under the agency of Johnson; one at Port St. Peter under the agency of Ed. Square; another in the same place under the agency of John LeMontais."[19] All of these firms would have to be dealt with if CRC was to become a going concern, for competition in the fishery was to be encountered not only in the marketplace but also at the production end. It was this duality of competition that made fish such a difficult staple to manage, requiring a complex of mercantile strategies in order to minimize the considerable uncertainties of the trade.

At the production end of the fishery, the nature of the resource itself was the root of the problem. As a mobile common-property resource,[20] the fishery was (and still is) theoretically open to all unless legal restrictions were imposed on fishing areas to control entry. In Gaspé, there was no entry problem, in the sense that proprietorial rights to the resource could not be established – fences could not be built round the fish stocks as they could round fields, nor could the equivalent of the "broad arrow" in the timber trade be devised.[21] This meant that exploitation of fish was not regulated, and so, in economic terms, the risk of dissipation of the economic rent was great.[22] In the case of CRC, what was really concerning Charles Robin was the potential loss of economic rent involved if local merchant competition became excessive, since that would reduce the amount of fish accruing to his firm (the number of fishermen being limited in these early years) and would, moreover, raise the possibility of price wars, forcing him to pay more for his bought fish in order to keep "his" fishermen from selling to other merchants. Both of these scenarios implied reduced profits, and they were further exacerbated by the ever-present risk of market glut. If too many firms sent fish to the same markets in the same period, the result was falling prices – and market prices for fish were notoriously unstable, being subject to political manipulation (in the form of bounties and tariffs)[23] as well as to the usual vagaries of the marketplace. While nothing could be done about competing firms outside of Robin's area of operations (such as those in Newfoundland, for example), at least within it he could control what amounts of fish went to market, and at what time of year, if he could engross the local trade – that is, create a monopsony. In other words, security in the fish business in these early years could be obtained only under monopsonistic conditions which dispensed with loss of rent, local price wars, and the dangers of local overproduction.[24] To establish such conditions in Gaspé, Robin had to find a way to establish informal property rights over the resource, and that was precisely what he was striving to achieve in the years between 1783 and 1793.

The problem of proprietorial rights in the Gulf of St Lawrence and the western North Atlantic fisheries has pervaded their history from the first exploitation of the resource up to the present day, and the issue has been resolved over time in a number of different ways. Access to the resource could be achieved either directly, through ships at sea which never touched shore (the French *"pêche errante"* or the modern factory ship), less directly through the establishment of shore bases from which the fishery was prosecuted (the English dry fishery which led to the settlement of the area), or indirectly, through the purchase of fish from residents who caught it them-selves but lacked the capacity to market it abroad. Whichever tech-nique was used, it was essential that access by others be restricted as much as possible, and ideally prohibited, if returns to capital were to be maximized and dissipation of the economic rent minimized. Hence the struggle between the French and the English in New-foundland; between English merchants and Newfoundland settlers; between merchants themselves in Newfoundland and the Gulf. This is why Harold Innis referred to the eastern Canadian cod fisheries as "inherently divisive" and why even today conflicts occur: pro-vincial against federal government, province against province, multi-national against independent producers, midshore against inshore fisherman. This is what really was at issue when Charles Robin worried about the presence of American, Nova Scotian, Québec, and other Jersey merchants in Gaspé; this is why he needed to capture "all the fish of the post."

In the late-eighteenth-century inshore fisheries of the Baie des Chaleurs, ownership of land brought with it the fishing rights of that land. It also carried the rights to timber which could be used for construction of vessels and of shore facilities – the physical ve-hicles for achieving access to the fish. Land ownership, then, had to be one of Robin's first concerns, since it was to be thought of in terms of the equation that he gave Fiott deGruchy: value of land equals control of the fish caught off its shores. In other words, it was not to be seen as a cost input in the normal sense of that term, so much as a means of access to the resource which the capitalist should secure for himself, thereby prohibiting competition from other would-be purchasers of the fish caught off its shores.[25]

Over time, the shore frontage of this land, with its stages and flakes for processing the fish, became a focus for settlement, as the labour force required to catch, dry, and cure the fish migrated into the Baie. In the early days, when labour was scarce, Robin dealt with it in the context of competition from other merchants, and either offered higher prices for fish (when there was no other option) or tied his fishermen to the firm through bonds of indebtedness. Lee quotes a leading Loyalist's complaint about this strategy, which involved local people in so much debt "as to oblige them to spend the whole Summer Season in fishing to pay up their arrears."[26] But as the population grew, labour became less scarce and – as the direct catchers of the staple – represented a potential new threat to the merchant's control of economic rent. Consequently, in order to keep his access to the resource secure, Robin found himself in the position of having to secure control of the labour force, not only to prevent other merchants from winning it away from him, but also to make sure that independent fishermen could not deplete the economic rent of the Gaspé fishery.

The strategies for labour control that Robin adopted varied according to the type of labour involved. These were of three kinds: those brought over from Jersey on a seasonal basis; those employed directly on the coast; and those transported seasonally into the area from other parts of eastern Canada.

JERSEY LABOUR BROUGHT TO THE GASPÉ COAST

Skilled labour was brought almost entirely from Jersey, especially in the early years, and the clerks in the firm were Jerseymen throughout the Company's operations.[27] The system involved shipping people into the area from Jersey at the beginning of the fishery season, and returning them to the island at season's end,[28] unless they were bound over for a longer period of service (customarily two summers and one winter for fishery shore crews; five years for clerks) to take care of premises and perform other activities, such as logging or bookkeeping, over the winter season.[29] This system had two major advantages: it relieved the chronic fear of unemployment in Jersey, and it provided nonresident labour for the fishery which, therefore, did not constitute a threat to Jersey control of the resource. It was, in effect, a far-flung form of maritime transhumance, involving as it did the employment of small farmers or farm-labourers from Jersey who could leave the care of their crops to their families during the spring and summer, and return home in the fall in time for harvesting.

More important, from the merchant's point of view, was the ease of controlling this kind of labour: indentures similar to those used for apprentice labour in the Old World could be employed. The New World indenture system has commonly been understood as a merchant-inspired mechanism for defraying transportation costs in the movement of labour from an Old World labour reservoir to areas in the New World where labour was a scarce (and therefore expensive) commodity,[30] and it has been argued that transportation costs for such labour would normally have been taken as accruing to such migrants, except that they were generally too poor to pay them. In the plantation economies of the Carolinas, for example, planters seeking a pool of cheap labour resolved this dilemma by paying the transportation costs themselves and then defraying the expense by indenturing this labour force to work for them for a set period of time upon arrival, free of charge. Similarly, and closer at hand to Gaspé, the Scottish settlement of Tracadie (the Glenalladale Settlement) used an indenture system to defray the emigration (transportation) costs of settlers emigrating from the Western Highlands of Scotland to Prince Edward Island. The indenture of one Alexander Macdonald in Morar, for example, reads thus:

It is contracted, indented and finally agreed between John Macdonald Esq[r] of Glenallandale and Donald Macdonald his brother german ... AND ON THE OTHER PART the said John and Donald Macdonald Bind and Oblige them ... to pay the charges of bringing the said Alexander Macdonald to the Island of St. John in North America AND LIKEWISE to pay him FOUR pounds sterling over and above his maintenance for each year ... Also at the end of said term to give him possession of Two Hundred Acres of Land which they hereby demise sell and to term lett ... from and after ... the expiry of this Indenture for the space of two thousand nine hundred and ninety six years.[31]

Writers on the subject of these indenture systems have therefore been able to equate distance from the Old World labour reservoir with cost to the planter; they have also noted that such transportation costs effectively prevented labour from migrating to the area independently of the planter class,[32] although in the Gulf of St Lawrence at least some emigrants (such as many in the post-1800 Highland Scots emigration) were independent.[33]

In the fishery, however, labour-transportation costs as such either did not exist or were minimal, since the ships involved were already making the voyage and were already a fixed capital input borne by the merchant. The only real additional cost was incurred on the return journey, when seasonal "servants" occupied freight space on Jersey-bound ships. In the West Country/Newfoundland fishery, passengers' fares appear to have been only rarely a negotiable item;[34] direct employees of ship owners normally did not pay, others ("passengers") did. But fishing-crew members who had their return passage paid were not infrequently abandoned or encouraged not to return, thus making a small profit for the ship owner.[35] Generally, passengers directly engaged by the West Country merchant firms to fish for their business were supposed to have their outward passage only paid.[36] The Jersey system appears to have been similar. There is no mention of transportation charges in the Jersey indentures I have examined,[37] nor do the letterbooks suggest that there were charges for indentured servants, although passengers did pay.[38] Nor does there seem to have been a parallel situation to the West Country experience of furtive leakage of the Jersey-imported force onto the coast. Labour did settle permanently in Gaspé – as can be seen from the surnames extant on the coast today – and good émigré planters were actively encouraged by CRC. Rarely, a letter will refer to the loss of a manager in an outport who had decided to set up on his own account,[39] and (also rarely) there is mention of fishermen avoiding the return voyage to Jersey,[40] but many of the crews and fishermen had engaged in the fishery as a seasonal additional income to that of a farm in Jersey, and had wives, families, and homes there to which they wished to return. The Jersey indenture system, then, cannot realistically be seen as an emigration device: its real purpose must be sought elsewhere.

Figure 3 shows the indenture of a Jersey clerk in 1845, and figure 4 is a transcript of the indenture of a Jersey sailor for the same year, the first indentured to CRC, the second to John Perrée. Neither mention transportation costs, and both contain punishment clauses designed to prevent the indentured servant from "leaving or quitting" before the expiration of the indenture. More significantly, both indentures clarify the benefits to merchant and servant alike, thereby allowing an assessment of the purpose of the indenture. The merchant, in effect, was provided with a source of cheap labour, as was true of the plantation or emigration indentures, but the servant (or "youngster") was in turn guaranteed not passage money, but his apprenticeship in a trade, in these cases as a clerk and a sailor, but

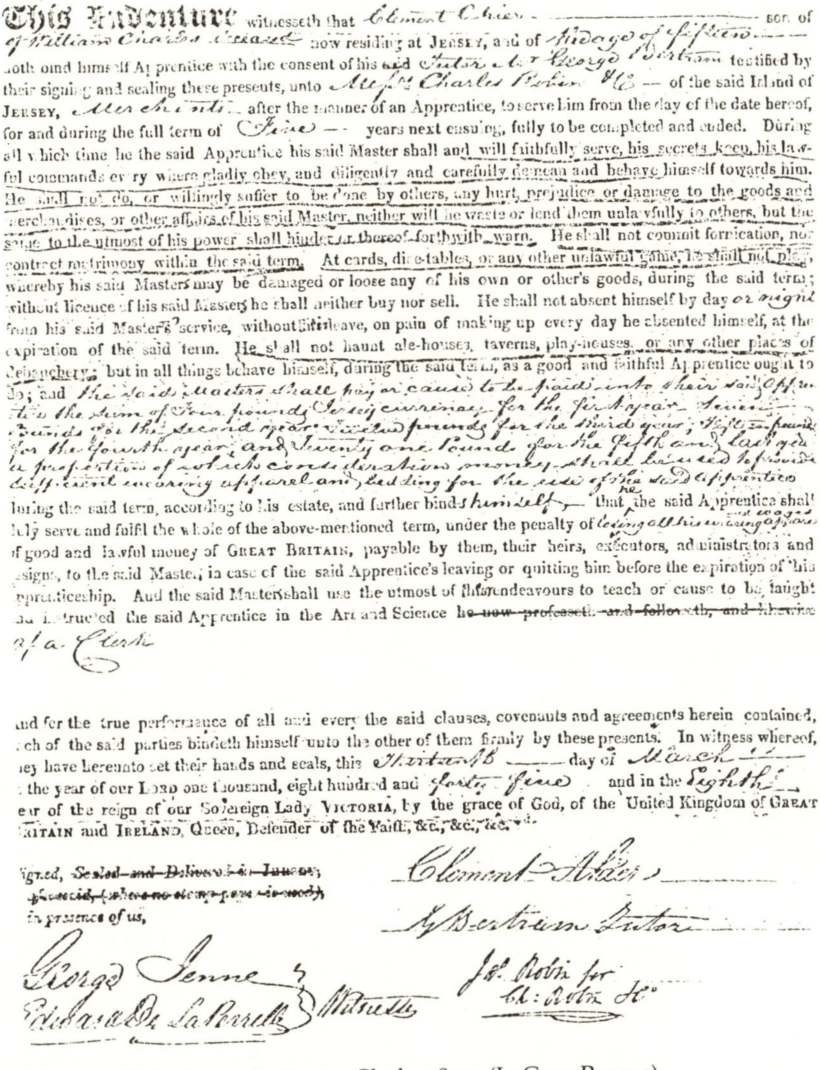
This Indenture witnesseth that Clement Ahier — son of William Charles Ahier deceased now residing at Jersey, and of the age of fifteen — doth bind himself Apprentice with the consent of his said Tutor Mr George Bertram testified by their signing and sealing these presents, unto Messrs Charles Robin & Co — of the said Island of Jersey, Merchants after the manner of an Apprentice, to serve him from the day of the date hereof, for and during the full term of Five — years next ensuing, fully to be completed and ended. During all which time he the said Apprentice his said Master shall and will faithfully serve, his secrets keep, his lawful commands every where gladly obey, and diligently and carefully demean and behave himself towards him. He shall not do, or willingly suffer to be done by others, any hurt, prejudice or damage to the goods and merchandises, or other affairs of his said Master, neither will he waste or lend them unlawfully to others, but the same to the utmost of his power shall hinder or thereof forthwith warn. He shall not commit fornication, nor contract matrimony within the said term. At cards, dice-tables, or any other unlawful game, he shall not play, whereby his said Masters may be damaged or loose any of his own or other's goods, during the said term; without licence of his said Masters he shall neither buy nor sell. He shall not absent himself by day or night from his said Master's service, without their leave, on pain of making up every day he absented himself, at the expiration of the said term. He shall not haunt ale-houses, taverns, play-houses, or any other places of debauchery; but in all things behave himself, during the said term, as a good and faithful Apprentice ought to do; and the said Masters shall pay or cause to be paid unto their said Apprentice the sum of Four pounds Jersey currency for the first year, [illegible] pounds for the second year, Twelve pounds for the third year, Fifteen pounds for the fourth year, and Twenty one Pounds for the fifth and last year a proportion of which consideration money shall be used to provide sufficient wearing apparel and bedding for the use of the said Apprentice during the said term, according to his estate, and further binds himself that he the said Apprentice shall truly serve and fulfil the whole of the above-mentioned term, under the penalty of losing all his wages and wearing apparel of good and lawful money of Great Britain, payable by them, their heirs, executors, administrators and assigns, to the said Master, in case of the said Apprentice's leaving or quitting him before the expiration of his apprenticeship. And the said Masters shall use the utmost of their endeavours to teach or cause to be taught and instructed the said Apprentice in the Art and Science ~~he now professeth and followeth, and likewise~~ of a Clerk

And for the true performance of all and every the said clauses, covenants and agreements herein contained, each of the said parties bindeth himself unto the other of them firmly by these presents. In witness whereof, they have hereunto set their hands and seals, this Thirteenth day of March in the year of our Lord one thousand, eight hundred and forty five and in the Eighth year of the reign of our Sovereign Lady Victoria, by the grace of God, of the United Kingdom of Great Britain and Ireland, Queen, Defender of the Faith, &c, &c, &c.

Signed, ~~Sealed and Delivered~~ in presence of us,

Clement Ahier

G. Bertram Tutor

George Jenne
Edouard De La Perrelle } Witnesses

Jas. Robin for Ch: Robin & Co

Figure 3. Indenture of a Jersey Clerk, 1845 (LeGros Papers)

frequently also as blacksmiths, carpenters, or skilled fishermen and shoremen. The purpose of the Jersey fishing indenture, then, can be said to have been that of defraying not transportation costs, but apprenticeship costs, which were carried by the merchant in return for a guaranteed cheap labour pool of men who were not emigrating but learning a trade at a distant, but still Jersey-owned, place of

Witnesseth that JAMES PAYN, with the consent and approbation of MR. PHILIP PAYN his father, bind himself apprentice unto MR. JOHN PERREE of the Island of Jersey, merchant, after the manner of an apprentice, to serve him, the said JOHN PERREE and other persons in the navigation of any vessel as the said JOHN PERREE shall order and appoint from the day of the date hereof for the full term of FIVE years and fully to be complete and ended. During all which said term the said apprentice shall and will faithfully serve the said JOHN PERREE and do and perform all such service and business as well at sea on board any ships and vessels which shall belong to, or be employed in the service of, the said JOHN PERREE....He shall not absent himself from the said service by day or night unlawfully....And the said Master his said apprentice shall and will cause to be taught in the art or business of a sailor as far as shall be necessary to the voyages wherein he shall be employed (there follows in handwriting:) and he shall furnish and allow unto the said apprentice a fully sufficiency of meat and drink and washing and further, the said Master shall cause to be paid unto the said apprentice the sum of £30 in the manner following, viz.
£4 at the expiration of the first year
£5 at the expiration of the second year

£6	third
£7	fourth
£8	fifth and last

And the said father obliges himself to provide and allow unto his said son good and sufficient clothing
etc., etc........
Signed: James Payn, Philip Payn, John Perree
Witnesses: Dan Dumaresq, Abraham Ahier.

Figure 4. Indenture (Perrée Papers)

training. Over time, then, what evolved was a colony of Jersey indentured servants (or *émigrés*, if they chose to remain on the coast), who were tied to Jersey through the merchants to whom they were indentured – a colony that therefore remained constantly aware of its roots. Indeed, even today Jerseymen express a continued affection for Gaspé: "La Côte! Pays tch'a un attrait et du sentiment tout spécial pour un vièr Jerriais comme mé ... Un pays tch'a veu ... l'industrie, les mînséthes – et la jouaie étout – d'la vie quotidienne dé tant d'Jerriais, nés en Jèrri, mais transportés sus la côte lé long d'ses grèves et au pid d'ses montagnes."[41]

LOCAL LABOUR ON THE GASPÉ COAST

In terms of control of the Jersey labour force, then, so long as it was indentured and operating within the terms of that indenture agreement, it posed no real threat to the merchant with respect to competition for access to the resource. When, however, any of the labour force settled, a different plan was required. This was the set of strategies evolved by the merchant-fishery entrepreneurs for control of the local population, and was commonly called the "truck system."[42] The system was not unique to Jersey merchants. It is well documented by Innis as having been on the rise in Newfoundland by the 1740s, as the population base increased and the supply trade to the island grew with it. At the same time, the basis for the old migratory fishery was being steadily eroded, and it is clear that many merchants in Newfoundland were shifting from the old migratory *modus operandi* to a more stable and lucrative supply business. There is no reason to suppose that the two functions were fulfilled by different firms; it is more likely that the shift to the supply trade was a progression in strategy by these same firms. The migratory fishery declined; the merchant who successfully made the transition from fisher to supplier did not.[43]

In Gaspé, Charles Robin started his business in the Baie des Chaleurs, trading salt and goods for fish and furs with the resident population in the 1760s, in order to attract fishermen to his company, and this general principle of exchange was to be the basis of CRC's trade over the whole century of its operations on the coast. Over time, however, CRC's exchange system became more complex, and as the coast grew more populous, the firm evolved three different ways to control this local labour. The simplest usually took the form of an *engagement*, or agreement, between CRC and a local fisherman. This generally functioned through payment in kind, or barter, with the cash element reduced to a minimum or zero. Typically, such a contract would read as follows:

I ___(Name)___, of ____(Place)____ do hereby willingly engage myself to Messrs Chas. Robin & Co. of _____(Place)_____ , to fish for them on their establishment at the said place by the Draft ... at prices which they may establish ... Should my account not be paid in full on the ____(Date)____ day of _____ , I will continue to Fish at the same terms till the a/ct. is paid ... It is well understood that I am to receive but the sum of Five pounds Barter ... When the Boat, Rigging, etc., is delivered to me it is all at my own risk till I deliver it at the expiration of this engagement. [44]

However, if the labour pool operated at some distance from Paspébiac and its outlying stations, as happened after the firm started to expand, a different method was applied. A case in point was that of the Magdalen Islanders who migrated each year to the Strait of Belle Isle or the Lower North Shore of Québec to fish for the Jerseymen. Here, a "floating" truck system was used to tie these distant workers to the merchant:

Les pêcheurs sont transportés aux endroits de pêche aux frais de la maison de commerce qui les engage; ils ont à leur disposition un bon bateau de pêche, parfaitement équipé ... et ils reçoivent pour chaque cent morues qu'ils livrent sur la tête du chafaud la somme de 5s 6d, moitié en marchandises et provisions ... Mais ils ont à se nourrir à leur frais, et si la pêche est peu abondante, leur compte de provisions, qu'ils laissent à leurs lignes, absorbent la plus grande partie de leur gain, et ils s'en reviennent très-souvent aux Îles de la Madeleine les mains vides. [45]

That is, the merchants tied these seasonal migrants by barter and debt not because they were a threat to merchant control, but because, like the Jersey seasonal fishermen, they were an additional means of access to the resource, and also because they were an additional source of profit through the sale of supplies.

The third form of control of the labour force on the coast was the truck system proper, which was in fact a trade-off between the interests of the individual fisherman and the merchant, albeit one that was heavily weighted in favour of the latter. In essence, it was set up so that instead of paying wages the merchant paid "truck" – that is, he provided the gear for the fishery, along with provisions on credit, against repayment in fish at the end of the season. This meant that the merchant carried the individual fisherman's risk of a season's capital outlay in gear and so on, while the fisherman in return guaranteed his catch – not his labour, but the results of his labour. Consequently, the merchant's control of access to the staple, albeit indirect, was strengthened, and his risk was kept minimal by the reduction of cash involvement. The price – that is, the value of supplies that the merchant would provide for a specified amount of fish – was set by the merchant, usually at the beginning of the season. In the case of CRC, such prices were usually decided in Jersey, ostensibly as a reflection of market prices for fish sold the previous season, and Paspébiac would be informed of the Jersey decision with the first boat to arrive in the spring. In 1788, for example, Charles Robin agreed to take one planter's fish at $8.00 per tierce because the previous year it made £1–17–6 at Leghorn less £0–8–0 for freight.[46] In years when other firms were operating on the coast, the prices offered by other merchants would also have to be taken into consideration when setting the price: "You will pay fish the same as last year's in barters, but for Cash you will have to keep up the price and not have Fauvel and LeBoutillier to consult together to fix our price ... There being a demand for Haddocks here you may pay them 10″;[47] or, again, "We are the only persons on the Coast who get it at 12/6 truck and 10/6 currency bill. Many pay it as high as 15/0 Bill which disappoints us of near 3,000 qtls."[48]

There is an extensive but impressionistic literature on the truck system in the Canadian fisheries.[49] How the system worked in Gaspé will be examined in chapter 5 in more detail. Here it is necessary only to recognize that it was important as a strategy for merchant control of local labour, and hence access to the resource (the catch). Its principal effect in the early years, when labour was scarce, was to tie the apparently independent fishermen to the merchant on a semi-permanent basis through bonds of indebtedness, while in the later years it served the function of facilitating business in a capital-scarce regional economy.[50] In essence, it assured the merchant of access to the fish, either directly (in the case of fishermen who were hired to fish for the merchant, as in the earlier cases discussed) or indirectly, through the catch of the "independent" fisherman who provisioned "on account" to the merchant. Such provisions were placed to the debit of the fisherman's account as "advances," and his catch, when sold to the merchant, was placed on credit to his account. One bad year in the fishery would be enough to ensure that a fisherman would complete a season either in debt or empty-handed. In either case, unable to raise the capital to provision himself over the winter months, he would have to seek advances from the merchant against his catch of the following season.

Abuse of the system by the merchant was obviously a simple matter, although there were certain built-in safeguards. The presence of other competing merchants on the coast in the early years ensured a variety of controls. One was that an individual merchant would only be able to lower his offering price for fish to the extent that he could still maintain his fishermen, since otherwise they would fish for a better-paying merchant and risk the legal consequences. As well, a good supply of provisions in the company store gave the well-stocked merchant a competitive edge: "I have now 17 boats of sharesmen engaged for Percé, all good fishermen, in fact the best in this place. Mr. LeBoutillier, having no provisions, has been the main cause of their leaving him. They all owe him, more or less, and are afraid that he will sue them. When I have more leisure I'll take deeds from those that owe us and in the amount include their advances."[51] However, merchants had to set their prices relatively within range of one another – although CRC does not seem to have "fixed" prices in an absolute sense[52] – since excessive inter-merchant competition on prices would have destroyed the whole basis on which the merchant system operated, by freeing the local fisherman from debt. This was what happened immediately after the American Revolution, for example, when New England vessels on the coast ventured to trade illicitly with the settlers, sending up howls of protest from the resident merchants.[53] Nonetheless, since traders into the area, while they might offer "cash on the barrelhead" to local fishermen, were not prepared to support them in bad years, these outsiders were really no more than irritants to a merchant whose truck system was securely established.

The local merchant also did not pursue price wars to the extent to which the fisherman could have established his independence, because of the inherent instability of the fishing industry in general. The fishery, like other resource-extraction industries, was susceptible to both resource depletion and market fluctuation, but it was especially unstable at the production end, since even without overfishing there were always bad years when the fish did not strike on the coast. A run of such bad years would mean a period in which the merchant's risk-carrying role of providing provisions and gear for local planters would work against him, since he would be forced to carry the capital costs without being able to realize his investment through the returns in fish that he would normally receive at the end of the fishing season. Fluctuation in the quantity of the resource extracted in any one year (or series of years) meant that sufficient capital resources had to be available to enable him to ride out poor market returns in the event of failure of the fishery. This is what had made it impossible for the fisherman to survive such a crisis; it was this investment risk that the merchant, with his larger capital resource, assumed in return for the fisherman's catch. Thus, traders were not a significant threat to the system. A merchant who over-invested (that is, provisioned unwisely) or who overpaid for fish, thereby depleting his capital, ran a grave risk of leaving himself with insufficient capital resources to see him through a series of bad years. Hence CRC's continual preoccupation with advances and constant concern that credit not be overextended nor the price of fish to the fisherman on the coast rise too high, even if this meant a loss of labour and therefore a contraction of CRC's access to the resource. The essence of the truck system, then, lay in the securing of control of the resource by the merchant; but the art of the truck system, in a competitive situation, lay in the fine judgment needed for its application, such that maximum control over the resource was maintained with minimum capital outflow.

It can therefore be argued that both the indenture and the truck system in the fishery were not merely "capital-saving devices: employees without families, long deferment of wages, truck pay and other ingenious credit mechanism,"[54] though they were all of these. They were also – and perhaps more importantly – the means by which the merchant minimized the risk of having his control of access to the resource challenged by independent indigenous fishermen. Jersey merchants' response to a resident population, then, took the form of moving one step backward in the production process. In other words, the resident fisherman now acquired the staple (although there were also Jersey crews fishing in Gaspé – that is, the more direct method was not abandoned), but he then had to sell it to the merchant in order to gain his supplies and, later in the year, his access to the staple market. Furthermore, since a growing population generated an increasing demand for supply goods into the fishery, the merchant increasingly took on the supplier function, payment being made by the fisherman in fish. The merchant also operated as a buffer for the local fisherman against the fluctuations of a frequently unstable market and those of an equally unstable resource, thereby gaining indirect control of the staple, his supply function being an additional benefit.

In fact, control of the fisherman as the means of access to the resource can be regarded as existing along a continuum: total control was needed where threat was greatest. The planter on the coast who fished for himself posed the greatest threat because he provided himself with his own equipment. He was consequently the most indebted to the merchant, since he passed this ownership and its cost over to the merchant. He was the proper object of the truck system, and it was his indebtedness that lost him his independence and tied him to the merchant, who thereby secured his loyalty and, more importantly, his supply account and his catch. A lesser degree of control (half cash, half barter) was exerted over the Magdalen Islander, who provided the merchant only with his labour while the merchant provided him with boats and gear, and gained an additional profit from provisioning him. This man was a fisherman only on a part-time basis; his real occupation was that of a farmer in a marginal environment, and the fishery provided him with an extra benefit in the form of additional employment. The least threat to the merchant was the man who was totally tied, the fisherman who could not provide his own equipment, and who was not paid even half in cash. This was the *engagé*, who sold his labour to the merchant at a fixed rate of barter. In other words, at one end of the continuum was the planter, who traded off his independence against security in the form of indebtedness; at the other end was the *engagé*, who was the fishery's equivalent of a landless labourer who could not hope to own the means of production: the fisherman without a boat.

MIGRANT SHOREMEN IN THE GASPÉ LABOUR FORCE

In light of the above description of the truck system, the manner in which the firms treated the seasonal migrant shoreman is instructive. There was a real difference between this seasonal contractual labour force and the previous two forms of labour. Contracts for migrant shoremen were usually very short – of perhaps five months' duration – and took the form of a fixed cash agreement. Such "servants" – splitters and salters, for example – were usually hired through an agent in such places as the counties on the South Shore downriver from Quebec City: "Si les saleurs Boulet, Nicholle et J. Bernaiche veulent retourner pour nous vous pouvez leurs accorder 1 ou 2 piastres par mois de plus que ce que j'ai specifié et autant au bon trancheurs qui ont déjà venus pour nous et qui vous connaissez bien."[55] It is significant that this French-Canadian labour, transported by the merchants from around Quebec City and employed seasonally on short-term contracts, was the only labour for which the Jerseymen paid completely in cash. This was also the only Canadian labour that was not resident in the fishing area and that did not actually catch the fish, but processed the catch on shore. It was the only Canadian labour, therefore, that did not pose any threat to Jersey hegemony in the Gulf. As such, it was not necessary to prevent capital formation among this group, nor in fact did the merchant have the necessary weapons for denying cash payment to them.[56] This does not, of course, mean that capital formation occurs only where cash is operative. The prerequisite for capital formation is that capital has to circulate and be free, and credits and debits can circulate freely, as can cash and bills. But this did not happen in Gaspé during the years of the merchant fishery because, for as long as there was a monopsonistic truck system operating on the coast, there was no true circulation of any kind.

The remaining problem that Charles Robin had to deal with at the production end of the fishery – controlling access to the marketplace – was solved in the process of controlling the labour force and dealing successfully with competition from other merchants. There was on the coast no means of transportation capable of taking fish to the range of markets required, or of handling the complexities of marketing the product, other than through the merchant system, and thus there was no other way that the staple could realize its value. Access to market hinged on the financial capacity to build or buy the sailing vessels capable of making the long-distance voyages involved, and this in turn demanded a relatively large capital input, such as that which existed only among the fish merchants of the coast.

The Perrée Papers give some idea of the cost of such a vessel on the coast in 1845: the brigantine *Chance*, built by Edward Mabé of Malbaie for Francis Perrée (See figure 5). The vessel, capable of carrying about 3,000 quintals of codfish, sold for about £560 Halifax currency, a price clearly beyond the income of a local fisherman, and one which represents the cheapest possible outlay for such a vessel since it was colonial-built, as the builder noted with some indignation during an exchange of letters between Malbaie and Jersey, written prior to the sale: "I suppose that you say, or any other gentleman might think, that from the statement of the price that the vessel must be very inferior built ... I consider that what we call the well-built vessel in this district is equal to the vessels of Europe."[57] The only possible further cut in costs that could be achieved was to build the vessel personally, and this CRC in fact did. The cost, however, would still remain prohibitive for the local fisherman, who has already been shown to have been unable to provide the capital for a year's fishing gear and provisions. The fisherman was therefore totally tied to the merchant in terms of market access.

In effect, there was a double issue of access and control on the coast: that of the merchant to the resource and that of the individual fisherman to the market. By 1793, Charles Robin had put in place a system by which the former was achieved. The latter could not be

```
Province of Canada
District of Gaspé
```

On this 9th day of June in the Year of Our Lord 1845, it is
mutually agreed upon between Edward Mabe of Malbaie, shipbuilder,
on the one part, and Francis Perree, of the Island of Jersey,
master mariner, on the other part.
That is to say:
the said Edward Mabe bargains and sells unto the said Francis
Perree a certain vessel now laying on the stock of Malbaie of about
160 tons of measurements, to be finished and completed in a
carpenter-like manner with all requisites spar complete cabin and
forecastle floor to be laid to furnish hardwood and timber for a
false keel and all the pine board or deal required for building or
making the cabin and forecastle.   To furnish 2 boats in all
defection timber to be taken out and replaced with good.   The
vessel to be caulked and sprayed and the old caulking done over
anew and made perfectly seaworthy to be ready for launching on or
before the 13th day of August next for and in consideration of the
sum of £3 - 10 - Halifax currency per ton old measurement, half
payable after launching by Bill on Quebec at 30 days sight and the
other half payable in like manner 1 year after, and 70 to be given
for the fulfillment of same or to be paid after launching at the
rate of £3 - 5- 0 Halifax cy per 10 old measurement.   The said
Edward Mabe further binds and obliges himself to pay to the said
Francis Perree the sum of 200 Hal. cy penalty on the event of the
nonfulfillment of this agreement.
The said Francis Perree on his part receipts of the foregoing
conditions and binds himself to comply strictly to the same.
Should any dispute arise respecting the workmanship or timber on
the said Vessel, the same shall be left to the decision of two
carpenters, one to be chosen by the said Edward Mabe, and the other
by the said Francis Perree.   The said Vessel to be launched and
delivered afloat at the sole risk and expense of the said Edward
Mabe and further the parties agree that the first mentioned price
or consideration shall be the condition of sale,
NAMELY

£3 - 10 - 0 per ton as aforesaid without interest
one half payable immediately after launching and the other half in
one year after.   Building Certificate to be granted at the time of
launching.

Done at Point St. Peter this day and year above written in the
presence of the undersigned witnesses.
                          Signed: Ed. Mabe
                                    Francis Perree
Witnessed: John LeBoutillier
```

Figure 5. Bill of Sale of the *Chance* (Perrée Papers)

achieved by the fisherman, because the necessary capital require-
ments could not be raised at the individual level locally, and there
was no alternative. Indeed, over the years CRC engrossed the mar-
keting for the whole coast, thereby effectively reducing other mer-
chants to middlemen for the Jersey firm. Samson, for example,
comments of Hyman and Company that they "had to go through
brokers in Jersey to get to the market. Channel Islands' shipowners
controlled all transportation and sales of cargoes." He adds that this
was the norm on the coast.[58]
```

Such marketing control at the production end was essential, since CRC could not adequately control the marketplace abroad. Markets for fish were diffuse, far-flung, variable, prone to glut, and, most importantly, outside the metropole. Thus, there was no relatively simple two-way exchange process between motherland and colony such as was characteristic of the early timber trade. The fish-trade merchant system was a three-way system, a triangle, which had to take the international arena into account. This required considerable skill and experience, well beyond the reach of the ordinary fisherman, and which the merchant developed over time.

In the early years of the business, CRC, alone among the various companies in Gaspé, managed to weather the difficulties of maintaining this system. Of the companies that had established along with it in the 1780s, all had failed by the turn of the century, and even CRC had extensive losses.[59] Robin's explanation of his competitors' failures was to note wryly that they had given up business "for want of success,"[60] arguing that Gaspé was too poor to support them.[61] What had actually happened, of course, was that only CRC had been skillful enough in its manipulation of the complexities of the production end of the trade and its handling of the markets to survive.

One example will suffice to show the entrepreneurial skill exercised by Charles Robin in order to maintain the business in its early years on the coast: that of the French Revolutionary Wars of 1794–1802, which produced the greatest crisis the firm would have to face until that which brought about its ultimate demise (see figure 6). Problems started in 1794, with the capture of St Sebastian in August and the first major threat to a Jersey fish market: "I ... conclude Balboa must have fell soon after at which rate the cargoes of the ships *Henry & Charles* and the *St. Peter* must have been lost."[62] By 1795 the position was worsening, and there was threat of war with Spain. Robin decided to send his vessels out anyway, provided there was no threat of a war with the United States, since, if Spain went to war, "we can always dispose of our fish in America where we would carry two cargoes for one to Spain ... I wish to make *Kingfisher* run thrice to New York and make considerable returns that way. My plan is not extensive and admits no curtailing at all provided we have Peace with America. The *Hilton* must be dispatched early from hence to Lisbon with a cargo of old and Fall fish for sake of a full cargo of good salt, which if war continues will command cash at a great advance."[63] In other words, with one market closed, alternative markets were still available and supplies brought to the coast would sell very well.

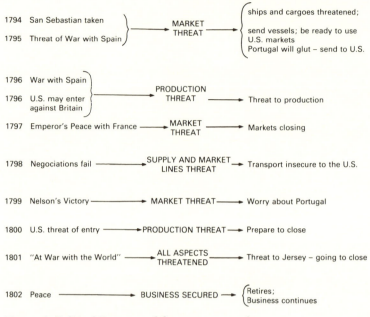

Figure 6. Political Events and Stresses, 1794–1802

By 1796, however, the situation was becoming more alarming for the firm. Robin feared that "we shall have, as once before, French, Dutch, Spaniards and Americans foul of us," and, as he could see no prospect for peace, he advised that "we should immediately get all our vessels insured for Lisbon where I think I shall order them all to touch."[64] Worse yet, he suspected the imminent entry of the United States into the war on the side of France. Then the firm would face not only the risks of losing markets and of dangerous transportation, but threats to its production bases as well:

I think if we could get our property insured against all risks, five months this year from 1 June to last of October and six months next year from 1 May to last of October, it would be a wise measure. It might be done now for a moderate premium and by and by it will be impossible. If the war continues it is but too likely we shall be involved in an American War. Whenever that is the case, then all we have here and at Cape Breton will be immediately destroyed, and you may depend on that as an undoubted certainty.[65]

By November, Spain had entered the war and markets were closing down. Robin foresaw a consequent glut in the Portuguese markets as "the only open market in Europe,"[66] and decided to turn to alternative markets: "I have given directions to Captain Hubert to procede [sic] ... for Salem or Boston ... and the cargo to be disposed of, and to procede in the same manner as the others to Portugal for salt."[67] The year 1797 saw the collapse of John Fiott and Company[68] and the likelihood of the Portuguese market's closing down. Robin wrote: "Had I known before of the Emperor's peace with France, I would have sent her [the *Kingfisher*] to the States. Now I expect Portugal will be under the necessity of shutting her ports to our Navy in order to obtain her peace. Thus Lisbon will be full of row-boat Privateers and there will be no access to it for our trade."[69] Seeing the European markets closing against him, he added, "Where to go now?" Indeed, not even the Jersey domestic market appeared to be safe, since he thought the French alliance meant the chance of an attack on Jersey.[70] The only solution was to have recourse to the United States markets – Boston, Salem, Cape Ann, Newbury-port, or Portland.[71] It was at this point, with threat to supply (Jersey), production (Gulf), and markets (Europe), that he decided to advise deposit of the year's proceeds in the United States.[72] In 1798, when negotiations for peace failed, the business was cornered at both market and supply ends in Europe, and Robin complained to his Quebec agents that "our business now is reduced to a very narrow channel and will not support itself if a change does not take place soon."[73]

At this point, luck entered the picture. The *St Lawrence*, bound for Boston with coal from North Sydney, "met near the coast of Cape Breton a Guadeloupe Privateer of 16 guns, which detained her four hours and liberated her without even plundering anything ... Thus you see, if we are unfortunate in some points, we are quite the reverse in others ... She proved old with a poor cargo which saved her, but according to the laws of war, she ought to have been de-stroyed."[74] But the message was still clear: CRC was now restricted and endangered in both markets and supply lines, and there were possible threats to both the home supply base in Jersey and the production bases in the Gulf. That fall, Robin shipped half of his fish to Boston, to go thence to Portugal only if he was unable to effect a sale in the United States. The other half was dispatched, under insurance, to Portugal, and he inquired about the time of departure of the convoys for Portugal and England from Newfound-land.[75]

In 1799, despite Nelson's victory at Aboukir Bay in the Battle of the Nile, which destroyed their seapower, the French continued to have great success by land. Commenting on French control of Naples, Robin expressed unease about Portugal, and about its precious market: "I'm afraid it will go like the rest and in a short time. Then we may hang up our hooks and lines, for in such a case I don't see what can be done with the fish. I'm afraid this is the last season we can carry on this business we have preserved this long time, but I see we shall be compelled to give it up."[76] To make matters worse, the *St Lawrence* was once again in trouble from privateers[77] – the supply and market sea lanes were clearly becoming increasingly dangerous. By year's end, the tide of war appeared to be on the turn, and Robin took time out to complain about British ships (sent to the Gulf to protect the fisheries) which had seized Philip Robin and Company vessels on a dubious charge of smuggling: "I am unacquainted with the system of condemnation, but it's evidently clear if His Majesty's ships which are sent on these stations with a view to protect the fisheries, will take every little mean advantage, instead of protecting, they will ruin them so effectually that they will finally [have] to go themselves on the fishery."[78] In 1800, the threat of United States intervention on behalf of the French increased once again, and Robin, already frustrated by danger-ridden supply lines and crippled markets, prepared to abandon the fishery. "It would be an excess of imprudence to continue business in this country ... if the prospect does not mend," he wrote,[79] adding that *if* the vessels were safe in Jersey, and *if* the United States remained friendly, and *if* he had a supply of fishing gear, then he would carry on "some business to the extent of 5000 qtls," which he would send to Quebec and Halifax "and then close the whole."[80]

The stress on the system, in other words, had finally become too great, and Robin had been pushed to the point of capitulation. Insecure markets in Europe, dangerous seas, the risk of sympathetic American ports "swarm[ing] with French privateers," and the possibility of the United States entering the war to support a flagging France meant that the Robin merchant system was threatened in its supply lines, its production centre, and its markets. CRC had survived threats to supply lines before, and to markets, but it could not tolerate an attack on its resource base.[81]

By May of 1801, with the threat of the loss of Portugal – the only remaining market – Robin declared that "now we are at war with all the World," and commenced making arrangements for the proper disposition of his personal assets. But Jersey appeared still to be in danger: "I do not know how anything can be secured, it's more risked in Jersey than anywhere else. I see our neighbours [France] are making great preparations. If they get it, they'll never part with it," he wrote.[82] He considered leaving his money with John Robin in Portugal as one possibility, provided John treated it "with proper caution: not to be all risked in trade," but he feared that Portugal might yet be conquered; he considered Canada, to the extent of about £2,000; and he considered Hamburg "now that the Emperor has made a peace ... I was made one of his subjects at Nieuport in 82." He advised his brother to "leave as much money in the States as you can."[83] By October 27th of that year, however, the crisis was nearly over, and Charles Robin was laying plans for the next year's fishery, taking precautions against possible future vagaries of war: "After next year we could put two of our five ships ... under neutral flag."[84] By January 4, 1802, he had news of the peace.[85]

The firm had survived; it had shown itself capable of taking stress in markets and supply lines simultaneously, and it had managed to carry on "a considerable business all the war."[86] It had outlived competitors and secured its foothold in Gaspé. In the succeeding decades, it would reach out over the Gulf to become a monument to Jersey merchant capitalism in the New World. When Charles Robin retired, in 1802, he left in his nephews' hands a business of which he could say that "a kind Providence ... has favoured us so far as to have made headway in a very poor country which satisfies me fully," although, characteristically, he would add, "I wish I was more grateful than I am."[87] But the traumas of the years from 1794 to 1800 had also pinpointed the weakness of the mercantile structure that he had created, a weakness which neither he nor his successors could ever really overcome. It was at the market apex of the merchant triangle, in the realm of international affairs, which was external to the fish trade and beyond the reach of its influence, that the industry was – and remained – vulnerable.

CHAPTER THREE

Organization at the Jersey Core

In this chapter, I will examine the management apex of the Jersey merchant triangle in terms of the establishment of power structures at the Jersey metropole and the creation of effective political strategies which were designed to enhance Jersey's ability to influence the larger United Kingdom metropole. This was necessary because the merchant triangle was vulnerable in various ways. One of these was its openness to mercantile and political power in the United Kingdom, within whose dictates the ownership and production apexes operated, and whose political power was strong enough to manipulate some of the vagaries of international politics which affected the marketing apex of the triangle. In creating their Chamber of Commerce, Jersey merchants were searching for stability in their major trade by seeking a way of communicating with external influences on that trade, a path through which they might be able to reduce their vulnerability. The operations of the Chamber provide a picture of Jersey mercantile interests at work at the ownership apex of the triangle as these interests sought to control as much as possible the larger world beyond the coasts of Jersey.

The Chamber's principal weapons were the traditional strategies of the island. It used the constitutional ambiguity of Jersey's relationship with Britain as a manoeuvring device with which to gain economic leverage or economic breathing space, and it sought to manipulate the mistrust of England for France by constant reminders both of Jersey's loyalty and its geographically advantageous position close to the coast of France in the not unlikely event of an outbreak of hostilities. With such an arsenal, the Chamber hoped to gain the full cloak of British protection, which was needed if Jersey merchants were to survive in the interstices of Anglo-French commercial power, in either Europe or the New World.

Table 1
Connections between Merchants in the Jersey Chamber of Commerce, 1768
(By Number of Connections[1] to All Merchants)

Rank	Name	W.P.	M.	L.	deG.	G.	LeM.	H.	D.	Hué	LeB	P.P	Pipon
1	W. Patriarche	x	1	1	2	2	2	2	–	2	3	1	2
2	Mallet	1	x	1	1	1	2	3	–	2	1	2	2
3	Lemprière	1	1	x	2	2	1	3	–	1	2	3	2
3	deGruchy	2	1	2	x	2	3	3	–	3	1	3	1
4	Gosset	2	1	2	2	x	3	3	–	3	2	3	1
5	LeMaistre	2	2	1	3	3	x	3	–	1	3	3	3
5	Hemery	2	3	3	3	3	3	x	–	3	3	1	3
5	Dolbel	–	–	–	–	–	–	–	x	–	–	–	–
6	Hué	2	2	1	3	3	1	3	–	x	2	3	3
6	LeBreton	3	1	2	1	2	3	3	–	2	x	2	2
6	P. Patriarche	1	2	3	3	3	3	1	–	3	2	x	2
6	T.&J. Pipon	2	2	2	1	1	3	3	–	3	2	2	x

Source: Minute Books, 1768.
Note: [1]Where 1 = direct connection; 2 = through one intervening link; 3 = two or three intervening links

To be able to put pressure on Britain to defend Jersey interests when necessary, the business sector of Jersey had to be able to present a unified front to the outside world. The merchant firms of the early Chamber (named in the "List of Signatories" written at the beginning of the first Minute Book) had in fact a substantial body of shared interests, as the Minute Book reveals. One typical set of membership entries reads:

Mr. Mallet	for Messrs. James Lemprière
Mr. Mallet	for Mr. Ab. Gosset, Matthew Gosset, deGruchy, LeBreton and Nich. Mallet
For my own a/c.,	Mr. Mallett

Entries such as this detail the shares held in vessels owned by groups of merchants: in this case shares for signatories 5, 6, and 7 in the vessels *Mary* (120 tons), *Defiance* (67 tons) and *Fortune* (37 tons), according to the list of vessels, owners, and signatory numbers given in the Minute Book. Since a member's name often appeared in combination with several different groups of signatories, as shown above, it is possible to see the various interconnections between members. These are shown in table 1, in which the first twelve merchants, who were most connected to all other merchants in the Chamber, are identified, along with their links to others among the first twelve. A "1" in the cell means that that merchant was directly linked to the corresponding merchant, a "2" that he was linked through another merchant, and a "3" that he was linked through less direct connections. For example, Pipon & Co. was directly linked to Gosset and to DeGruchy, and through them to Patriarche, Mallet, Lemprière, LeBreton, and the other Patriarche, then through these connections to LeMaistre, Hemery, and Hué.

This pattern of shared interests among the merchants of Jersey was not unique to the early years of the New World trade. In the nineteenth century, they remained "prone to organizing their fishery by means of societies with multiple and cross-fertilized partnerships,"[1] as figures 7 and 8 show for the firms of Robin, Janvrin, DeQuetteville, Nicolle, and Gosset. (The pattern would be even more complex if eighteenth-century links, such as those shown in table 1, were also included.) A substantial part of these networks can be explained by examining the family and marriage ties – sometimes formed prior to business partnership, sometimes after partnerships had already come into existence – of the Jersey firms. The Robin family is a case in point. In 1840, PRC (Philip Robin and Company) was made up of James Robin, Clement Hemery (see figure 8: he was James Robin's father-in-law, and also father-in-law of Philip Janvrin), Philip Robin, Elizabeth Robin, and John Robin. In that same year CRC was made up of James Robin, Francis Janvrin (and later his heir Frederick Janvrin), Philip Robin, Elizabeth Robin, John Robin of Liverpool, Frederick DeLisle of London, Thomas Pipon of Surrey, and Philip Raoul Lemprière of Jersey. The last two were part of Robin, Pipon and Company, for which firm Charles Robin had been acting as agent when he first went to Gaspé. The Janvrin and DeLisle marriage connections were made in the second generation of the firm of CRC. John Robin married Mary King of Liverpool; John was the Liverpool agent (Robin and King) for CRC. James Robin's sons were Raulin Robin and C.W. Robin. Raulin married the daughter of the Naples agents for the firm (Maingay); C.W. Robin married Elizabeth Janvrin, and went into business with Isaac Gosset, who married C.W. Robin's aunt Madeleine, who was James Robin's sister, James in turn being married to Isaac Gosset's sister. Robin Frères was the firm founded by C.W. Robin's sons, who were Janvrins on their mother's side. By 1840, the firm of P. and F. Janvrin comprised Francis Janvrin, Jr, and Frederick, his son. Francis's sister Elizabeth married Frederick DeLisle (London agent for Janvrin and Robin), and their daughter married Thomas Grassie, Halifax agent for CRC and PRC. Elizabeth Janvrin, daughter of John Lewis Janvrin and Julia Durell, married C.W. Robin, and it was her sons (Charles Janvrin,

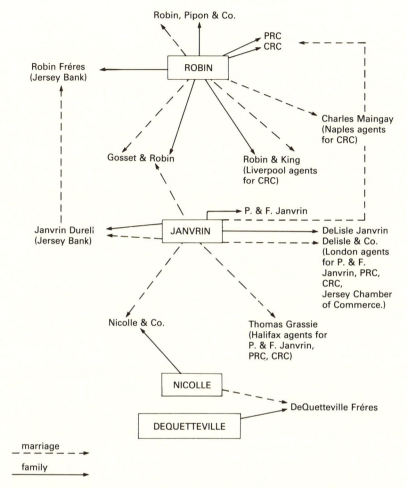

Robin, Pipon & Co.

PRC
CRC

ROBIN

Robin Fréres
(Jersey Bank)

Charles Maingay
(Naples agents
for CRC)

Gosset & Robin

Robin & King
(Liverpool agents
for CRC)

P. & F. Janvrin

JANVRIN

Janvrin Durell
(Jersey Bank)

DeLisle Janvrin
Delisle & Co.
(London agents
for P. & F.
Janvrin, PRC,
CRC,
Jersey Chamber
of Commerce.)

Nicolle & Co.

Thomas Grassie
(Halifax agents for
P. & F. Janvrin,
PRC, CRC)

NICOLLE

DeQuetteville Fréres

DEQUETTEVILLE

marriage

family

Source: Robin and Janvrin genealogies, Ships' Registrars, Letterbooks, Family Papers

Figure 7. Linked Directorships; Kinship and Marriage Ties, 1840

Philip, and Snowden) who took over the banking firm of Janvrin, Durell and Company, at which time it became Robin Frères: the Commercial Bank of Jersey.

Figures 7 and 8 also show that this network of cod-trade families gained a measure of supply and market area control, as well as financial control, through marriage ties. The connections back to families given in the initial list of subscribers to the Chamber of Commerce (table 1) suggest that the same process of marriage links

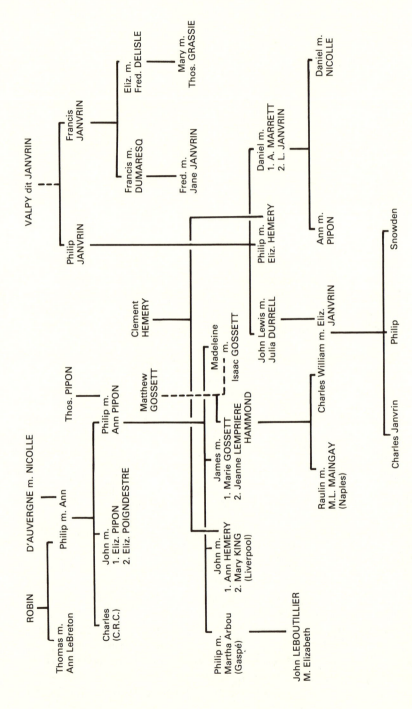

Figure 8. Robin and Janvrin Genealogy

was operating in the eighteenth century as was found for the nineteenth century. The names Robin and Janvrin represented minor shareholders in 1768. By 1840, they had become among the most important of the cod-trade firms, having established an extension of control of the trade beyond Jersey and Gaspé firms into the Jersey banking system, into the Halifax market system and supply structure, and into the Mediterranean marketing structure. In the case of supplying and marketing, additional information flows were created. In banking, more direct part-ownership was achieved. This nineteenth-century control was established, as the eighteenth-century links had been, through connections with leading families, including such people as Frederick Janvrin and Isaac Hilgrove Gosset. In this way, control was extended beyond CRC into the firm of Nicolle et Cie, and from that firm into DeQuetteville Frères.

A closer look at the evolution of the Robin/Pipon, Janvrin, and DeQuetteville firms shows that, despite their interrelatedness, they were nonetheless distinct from one another in some ways, and indeed represented different paths in the development of Jersey merchant capital. The Janvrins, like many of the "wealthy merchants of St Aubin, had graduated from sea-captain to owner to 'armateur.'"[2] From early on they were involved in both local and international trade, accumulating capital derived from local coasting, the American tobacco trade, and possibly smuggling as well. One of the earliest firms in the Newfoundland fishery, they further advanced their business with privateering until, by the 1750–90 period, they were involved in the cod trade, the local trade with Britain, and the supply trade to Newfoundland out of Waterford, and were also part-owners of nine vessels. Along with some other firms, they branched out into the Gulf fishery and, in partnership with the Robins, owned a fishery in the Magdalen Islands.[3] By 1816, their *modus operandi* was changing – "On January 13th M.D. Janvrin's stores situated in Pier St are to be let in vacant possession, and Messrs P. & F. Janvrin & Co. will put up for sale on January 26th le Schooner Angelique"[4] – they were shifting, it seems, from being merchant capitalists to being finance capitalists; in 1817 they appeared in the Jersey almanacs as "Janvrin, Durell, DeVeulle et Cie" – the Commercial Bank. Thereafter, they gradually sold their shipping interests, until by 1850 they retained only three vessels. At this point they also sold their Gaspé business at Grand Grève, and by 1860 they no longer had any shipping interests in the island.[5]

The DeQuetteville firm followed a different pattern. In the 1720s, they were a family of local status in Jersey, owning three vessels, and were involved in the Barbados and Mediterranean trades. During the 1750–90 period of expansion in the Newfoundland fishery, they were established in Harbour Grace, moving from there to La Poile and up the west coast of Newfoundland to Labrador as the old fishing centres in the Avalon Peninsula grew more thickly settled. Thereafter, they conducted their principal operations from the Strait of Belle Isle, and by 1857 – with 150 men and fifty boats at Blanc Sablon and establishments at Forteau and Isle à Bois – they were the biggest concern on that coast.[6] Philip DeQuetteville became president of the Chamber of Commerce, and continued to operate his business in the traditional manner with control always exerted from Jersey, although by the 1850s other firms on the Labrador coast of the Strait of Belle Isle had shifted from being fishing-ship operations to being resident and St John's-based enterprises.[7] It is possible that this lack of flexibility was a major cause of the eventual downfall of the firm. Its continuing dependence on Jersey left it open to the vicissitudes of Jersey finance (which was far from stable at the time), and resulted in the collapse of the firm in the bank crash of 1873, just as Slade's of Newfoundland had collapsed a decade earlier in the wake of the Poole bank crash of 1861.[8] The Jersey bank crash, however, may have been no more than a partial cause of DeQuetteville's failure. It is quite likely that the firm managed to survive as long as it did only by virtue of its location in Labrador – the Strait of Belle Isle remained at the frontier stage of development for a long time, and so the continuance of a Jersey-based ship fishery was possible there.[9] In 1863, however, as the economy of the area began to mature, legislation to prevent the importation of supplies directly from places outside Newfoundland was instituted.[10] DeQuetteville, as president of the Jersey Chamber of Commerce, protested to Parliament in a petition dated September 9th, 1863, which clearly demonstrated his old-fashioned perception of this fishery. By the 1870s, all other firms in the area had been taken over by Job Brothers of St John's and Hants Harbour.[11] DeQuetteville, continuing as before, failed in 1873.

Pipon, the third of these firms, started as the agent for another Jersey firm at Dartmouth. By the 1720–50 period, with ownership in nine vessels, it was in the Jersey import trade and the cod trade; its partnership with Fiott must have occurred at about this time. The firm followed an expanding Jersey cod trade into the Gulf of St Lawrence and established itself at Arichat, with the Robin family. From this partnership came the firm CRC, in which the Pipon family remained co-partners.[12]

Figure 9 shows the structural aspect of the evolution of firms involved in the Jersey cod trade. The necessary expertise, motiva-

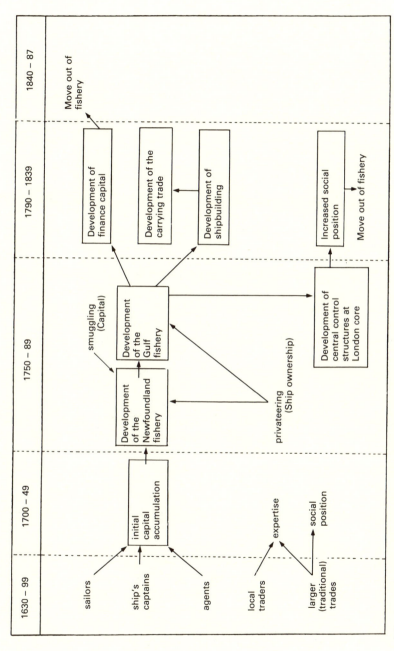

Figure 9. Structural Aspects of Jersey Firm Evolution, 1630–1887

tion, and capital for entering the trade were established by these firms in the centuries preceding the development of the Newfoundland fishery by Jersey. Their interests evolved naturally through trades, businesses, or occupations which gave them expertise in the kinds of skills the cod trade demanded. The Janvrins, for example, were merchants and sailors, used to international trading especially in the New World, who acquired through the generations the requisite capital for the formation of a company involved in ship-owning and the cod trade. They also acquired the expertise needed to move from this trade into the London commission business and into banking. The DeQuetteville firm grew out of trading in wine and fruit in the Mediterranean – a trade in which it was still involved in 1830. The Pipons were initially agents in Dartmouth, the port from which so much of the Newfoundland trade was organized. They ultimately (in their later guise of CRC) became the most successful of the cod-trade firms.

There was one firm that had common bonds with the three others outlined here but had a different function in the Chamber of Commerce. This was the London firm of Fiott DeGruchy, used as a commission agent by Pipon, DeQuetteville, and Janvrin. The Chamber recognized that the economic "imbalance of power" was firmly tilted toward London and that the Jersey merchants required the services of a metropolitan firm to represent their mercantile interests at the London core. Fiott DeGruchy was selected to do this job. When it operated as an agent for individual merchants, Fiott DeGruchy dealt with trade in the sense of specific transactions in the marketplace; as agent for the Chamber it acted as a watchdog over impending legislation in the British parliament which could affect Jersey trade in general for good or ill. Its job was to make representation at the various legislative arms of the British Crown when the need arose, and here it was aided by other various connections of the Chamber, such as the lieutenant-governor and other prominent Jerseymen with influence at the British core.

As a metropolitan firm, Fiott DeGruchy demonstrated a rather different mode of mercantilist development than the ordinary cod-trade firm. The family evolved out of shipping: "All young men who were followers of Admiral Durell and my uncle then went into his father's own ships. He commanded one to Newfoundland before he was seventeen," reminisced a member of the family in 1771.[13] By 1772, the firm had made its move away from the simple cod-trade business to establish commission agencies in London, in Paimpol, in St Malo, and in Jersey.[14] In 1774, John Fiott in London (who had until this point represented his father, the Seigneur of Melèches and a Jersey resident)[15] went into partnership with Philip DeGruchy under the name of DeGruchy and Fiott. The firm prospered, commanding much of the Jersey merchant business.[16] By 1777, John Fiott's share of the profits, exclusive of occasional business, was £1,900,[17] and by 1781 he had hopes of marrying into the English aristocracy, via the hand of an heiress by the name of Harriet Lee, ward of Lord Chief Justice Lee – Sir William Lee (Bart.) of Hartwell.[18] John Fiott's letters of this time to Sir William provide a comprehensive picture of the business, and of his personal fortune (which he claimed was equal to that of his chosen bride, who was heir to £10,000): he made upwards of £1,000 per year from commission alone for his share of the business. Fiott described the firm in 1781 as being an "old established Jersey House"[19] started by Henry Durell and continued by James Pipon until he was succeeded by DeGruchy. The major business conducted by the firm in the Channel Islands was a "business on Commission,"[20] consisting of the purchase of ships and goods "by order of our different Correspondents, to be consigned as they direct." The firm also sold ships and goods consigned by these correspondents "for their accounts," and carried out day-to-day financial transactions such as insurance, stock sale and purchase, bill payment and receipt, and so on. Fiott claimed that the firm's social connections with the leading families in Jersey gave him "the chief of the business" of the island, and added that the firm was joint shareholder with many of its clients in their ships and commodity purchases.

The firm was not confined to the Channel Islands, or London, but also had important connections with Norway, where business was conducted in the amount of £5,000 annually with Lady Ancher and Sons of Kristiania, the chief merchant house in Norway, according to Fiott. This business consisted of "receiving their bills for the amount of all their cargoes of deals, masts, balks, iron, tar, etc., which they ship to the different ports of England, Scotland and Ireland, making their insurances and receiving consignments of such cargoes as they send to London, 8 cargoes of which article we have received and sold for them this year."[21] Indeed, the firm was expanding this Scandinavian interest and acquired the function of relaying to the Danish Ambassador each year his appointment and salary from the Court of Denmark. Fiott concluded:

Our commission business alone, exclusive of the concerns we occasionally take in ships and goods with our Friends, has for these 3 years past amounted to £2000 per annum. This year it has exceeded it ... Our Friends have promised to obtain for us a share of the Danish East India Company

and of the Royal Bank Counting House at Copenhagen ... My fortune is
only about £5000, but it is chiefly the produce of my business ... This is the
true state of my business, my connections and my fortune. Upon a consideration it will plainly appear to be the most clear, certain and safe business
that can possibly be, and by no means of so precarious a nature as that of
merchants whose sole business is in adventures on their own account.[22]

Unfortunately, his predictions proved wrong. His partner of later
years, Philip Gavey, mismanaged the funds, and in 1795 the partnership dissolved. John Fiott died at Bath two years later.

Fiott DeGruchy, then, as John Fiott explained, evolved from a
family with a sailing tradition to the command and eventual ownership of sailing vessels. The path this firm chose led it into the
commission business, with diversified interests that went well beyond the cod trade by the 1780s, in a classic pattern of eighteenth-
century merchant-capital expansion right down to the search for
status, land, and position. Fiott DeGruchy most probably failed from
over-extension (as opposed to DeQuetteville's failure, which could
be argued to have been insufficient extension), if that is how Philip
Gavey's mismanagement of funds is to be interpreted.

Considering these four firms together, DeQuetteville was the most
traditional, Janvrin the least so, in that it made the shift to finance
capital. Fiott can be argued to have been the most evolved in the
sense that its owners progressed most completely through the
evolutionary process from sailor in the first generation to marriage
into the aristocracy in the last generation, while Pipon was the most
successful, in that it evolved into the dominant cod business in the
Gulf, adapting and diversifying throughout its long period of existence, until the differing requirements of another age forced its final
collapse. Each firm was characterized by a distinct approach to the
cod trade, and each flourished at a particular stage in the evolution
of Jersey merchant capital. Together, they represent the kinds of
cod-trade merchants who were involved in the Chamber.[23]

How, then, did the Chamber, as a collectivity of these and other
merchants like them, function in its capacity as representative of,
and advocate for, their trading interests in the larger world of the
British Empire and its trading system? During the years of Jersey
merchants' involvement in the cod trade, the Minute Books of the
Chamber offer an ongoing account of all the commerce of the island
and the state of its economic health. Up to 1870, the major interest
of the Chamber lay in the cod trade as the most significant feature
of the island's prosperity and therefore one to be protected at all
costs. In 1796, for example, the Chamber sent a memorial to the

Right Honourable Lord Granville, "one of His Majesty's principal Secretaries of State," stating that "the cod fishery on the coast of Newfoundland and parts adjacent is the principal trade carried on from this Island, employing in times of peace between 60 and 70 vessels and about 2,000 seamen ... The British fisheries on the Isle of Newfoundland and parts adjacent ... would alone be more than sufficient to supply the markets of Spain, Portugal and Italy"[24] – they begged Lord Granville, therefore, not to allow renewed French competition in those fisheries when the peace treaties following the Napoleonic wars were negotiated.

Such memorials were usually presented not only to the relevant authority in London, but also to the States of Jersey, thereby ensuring maximum pressure on the British parliament to legislate in favour of Jersey interests. In 1805, for example, while seeking to prevent the passing of a bill before Westminster which would prohibit the importation of "not only spirits and tobacco but also ... every article that pays duty in England," the Chamber sent a memorial to the States noting

... that your Memorialists are justly alarmed at the great crisis which menaces our very existence. They trust as guardians of the rights and privileges of the inhabitants of the Island you will adopt such measures ... as your wisdom shall seem meet, as it appears to your Memorialists that if the Bill now pending before Parliament was carried into effect it would so far operate as to crash the almost only branch of trade which your Memorialists enjoy, by salt not being allowed to be brought in vessels that are not upwards of 200 tons ... Your Memorialists think that this is a fair amount to demand that the bounty allowed in merchantable dry cod fish imported into England from Newfoundland, etc., should be extended to this Island in other vessels than those which have brought it direct from Newfoundland aforesaid for the purpose of being sent for a market to some part of the United Kingdom.[25]

Patronage was a vital component of the Chamber's strategy. A lieutenant governor who was concerned for Jersey's well-being was an invaluable asset in times of strain between London and island interests. In 1786, for example, the Chamber of Commerce "resolved unanimously that the thanks of the Chamber be given to the Right Honourable General Conway, Governor of the Island of Jersey, for his parental care and influence in promoting the trade and navigation of this island, for the commercial privileges already obtained through his means, and the prospect of advantage entertained under his protection."[26] In November of 1786, the Chamber of Commerce took note of the fact that the articles of the treaty of commerce just drawn up between England and France contained clauses which meant that Jersey was not allowed "the same liberty of importing all French manufacture into England as we hitherto enjoyed with respect to the Articles of wines, etc.," and resolved that they should "beg the Lieutenant Governor to write the Lord Beauchamp to desire his assistance in Parliament to obtain the privilege of importing French goods from this Island as if directly from France."[27] This is a typical example of how the actual wielding of power worked in relation to Jersey, the dependence of Jerseymen on personal contacts with powerful British peers being a continuing feature of political life on the island. It is also a good example of Jerseymen's search for a secure foothold in the realm of Anglo-French trade.

The power of London was crucial, even at the individual level. In wartime, for example, Jersey merchants could save themselves large outputs of capital for ships if they became involved in privateering, but in order to so do they had to acquire a letter of marque from the Honourable Board of the Custom House in London. In 1806, for example, John Janvrin applied for a letter of marque: "Mr. John Janvrin Owner of the cutter Providence Ph. Hammond Master having acquainted me that he has apply'd for letters of Marque for said Vessel. I beg leave to observe to you for the Honourable Boards information that said Owner is a respectable Merchant of this Island and that to the best of my knowledge and belief that neither him [sic] or the Master has been concerned in smuggling."[28]

Decisions that were of vital importance to the entire cod trade (on matters such as bounties and tariffs, for example) were likewise made in London. In a letter to the Honourable Board. on the 19th of August, 1806, it was stated that Philip Nicolle had applied on behalf of "himself and others of this Island" concerning the bounty on salt fish and salmon imported on the 14th of August 1806 from Fortune Bay, Newfoundland, having presented the text passed on the 16th "for allowing until the 1st day August 1807 the importation of certain Fish from Bounty thereon."[29] In times of severe hardship, the members of the Chamber did not hesitate to petition the monarchy itself. In 1816, they sent a petition to the Prince Regent, forwarding it through the lieutenant governor, whom they requested to give the petition "a favourable recommendation ... to the consideration of His Majesty's Ministers":

For many years the inhabitants of this Island derived great advantage from the fisheries from Newfoundland but owing to the present competition of the French, Americans and Danes that trade is become ruinous, is nearly annihilated, and several hundred seamen are thereby deprived of the means

of subsisting themselves and their numerous families. That by certain restrictions on the trade of the Islands executed by Parliament in 1805, and
45 George III c. 121 number 3, the trade with France is in some measure
restricted ... [30]

The Chamber's petition, and the accompanying letter from the president of the Chamber, went on to elaborate that "every possible
channel of trade and industry is closed" and emphasized that if some
help was not given, then "emigration to a great extent must take
place and numbers of the inhabitants of this Island will be bereaved
to seek in foreign countries those resources and means of subsistence
which they can no longer find in their mother country."[31]

Gradually, the Chamber gained a solid legal basis for the better
prosecution of the fisheries. During the years from about 1835 to
1859, its requests were viewed with increasing favour in London,
although always within the framework of Jersey's ambivalent constitutional position with respect to the legislative authority of Britain.
In 1835, for example, the Privy Council for Trade responded to
representations made by the Chamber for the removal of some duties
levied on British manufactured goods imported into British North
America from Jersey, as follows: "It does not appear that the position
of the merchants of Guernsey and Jersey has been altered by any
recent law, although it may have happened that the existing law
has lately been executed with rather more strictness than has hitherto
prevailed."[32] An accompanying letter from the Treasury Chambers
added:

The Channel Islands have never enjoyed the colonial trade upon an equal
footing with the United Kingdom, and My Lords do not feel disposed to
authorise the conception of such an equality. But in the intercourse between
the inhabitants of Jersey and Guernsey and the fisheries of Newfoundland,
Labrador and the northern coasts of America, My Lords are of the opinion
that the utmost indulgence which a liberal construction of the law can permit
should be granted, and that the freedom from duty conceded by the law
on the importation of goods for the use of the fisheries, should be held to
include all articles which can be useful to those inhabitants, except such as
are obviously those of luxury and refinement.[33]

And with that the Chamber of Commerce had to be satisfied.

In 1841–42, the Chamber exerted its influence outside the British
legislative process. Spain had placed a tariff on fish entering its ports
in any manner other than directly from the fisheries, and Jersey
merchants found themselves unable to bring fish to Spain via Jersey.

The Chamber wrote to the Spanish consul in London that the merchants involved in the British North American fisheries were complaining that the tariff meant that the expenses incurred left them unable to compete in the Spanish market:

This Island has extensive fishing establishments at Newfoundland, Gaspé, and Labrador, and ... several cargoes are landed here during the Fall of the year with the intention of sending them during Lent either to Spain or some other market in the Mediterranean. There are beside several other cargoes sent direct from the fisheries, but the fish that has now been landed here with the intention of being trans-shipped to Spain can no longer be so, the extra duty being nearly tantamount to a prohibition.[34]

The result, the Chamber pointed out, would be that merchants would have to send this fish to other, glutted, markets, leaving Spain with an inadequate supply for its needs. A further difficulty lay in the need to get fishermen and landsmen back to Jersey at the end of each season; to take these passenger vessels via Spain would be a considerable additional expense and inconvenience for the merchant. The letter to the consul then pointed out that not only would Jersey be adversely affected by the tariff, but Spain also, since "one of our merchants has now in store a cargo of codfish which he intends sending by one of his vessels to Valencia, and bringing back a ported cargo of wine, brandy and fruit. But by the new law he is debarred from this speculation, and is in some respect driven to import from other countries the same kind of goods which he would have purchased in Spain. I could mention other parcels of codfish here originally destined for Spain which must be sent to other markets."[35]

At the same time as Charles LeQuesne was writing these carefully couched threats on behalf of the Chamber, Philip DeQuetteville, as president of the Chamber, was writing to the president of the (British) Board of Trade and to H.M. Secretary of State for the Foreign Department, seeking assistance through the diplomatic channels they commanded. His letter provides a picture of the cod fishery at mid-century:

The principal commerce of the Island, my Lord, that which employs a great part of its shipping and a great number of its inhabitants ... are the fisheries at Gaspé, Nova Scotia, Newfoundland and Labrador. There are employed ... as seamen, fishermen and landsmen about 4,000 persons. There are in this Island many families engaged in the making of worsted hose and mitts, wearing apparel, boots, and shoes for the use of the fisheries. The trade gives employment to about 8,000 tons of shipping, exclusive of those vessels which carry fish to the Brazilian and other markets from this Island. It is therefore with peculiar interest that the Chamber of Commerce regards this branch of commerce and whatever is favourable or injurious to it. That trade destroyed, that source of industry dried up, the commerce of this Island would receive a death-wound, and many a parent who now, by the industry which this trade supplies, maintains his family in decent pride and honest independence, would probably be obliged to seek relief from his parish, which is a degradation and a stain which a Jerseyman can never forget or overcome.[36]

DeQuetteville then proceeded to set the Spanish tariff within the context of a restriction on the British fishery and on the commerce of Britain, and requested that a modification of the Spanish law therefore be sought by the British government. In the dawn of a British policy of laissez-faire, he pointed out that

... in a spirit of friendship and reciprocity, the Spanish government should not act with that severity towards British commerce, for a great part of the salt used in the curing of codfish at the fisheries at British North America is imported there from Cadiz. If the chamber were desirous that a similar system should be observed towards Spain as it manifests against us, they would pray Your Lordship to induce the British Government to levy an export duty of the same amount on all codfish sold to foreigners at the fisheries, or shipped in foreign bottoms; but they are more desirous of seeing a friendly feeling and friendly legislation between nations in commercial affairs, and would much rather that foreign governments should act towards us in a liberal spirit than that we should, as a return of hostility, feel compelled to enact laws injurious to their commerce and prosperity.[37]

Having thus proclaimed himself firmly in the mainstream of liberal British thinking, DeQuetteville left the matter in the hands of His Lordship. The results of this, possibly the most effective effort in economic diplomacy carried out by the Chamber, were not long in coming. In a letter written from London on the 17th of March, 1842, the Spanish consul explained that "the circumstances of having touched at a national or foreign port or ports previous to entering the one in which the cargo shall be discharged shall not subject it to any other duties than that to be levied on codfish imported direct, it being however necessary that sufficient proof be afforded to the Custom House of its shipment at Newfoundland."[38]

The year 1860, however, proved to be a turning point in the success of the Chamber in promoting both the prosperity of the cod trade and Jersey's general trade and navigation. From 1835 until this time, London was favourably disposed toward appeals by the Chamber and lenient in the interpretation of British legislation as it affected the island. But the weapon of Jersey's ambivalent constitutional position with respect to the United Kingdom was a double-edged sword, and Britain could always exert its power in a way that was detrimental to Jersey interests. In 1860, the storm clouds began to gather. On the 12th of June, the president of the Chamber wrote to the commissioners of Her Majesty's Treasury to protest Jersey's inclusion in the Customs Amendment Act, which he claimed laid charges upon goods exported from foreign parts or the colonies into Britain, and from Britain into these foreign parts or colonies. He protested vigorously that Jersey was *not* foreign, and that even if the act intended to include the colonies, Jersey was nonetheless still exempt: "Your Lordships are no doubt aware that the Channel Islands have peculiar privileges which have been admitted and acted upon by the Parliament and Her Majesty's government and which have been secured to them by the express provisions of these charters to give to the people the amplest immunities and privileges in all English ports and markets, free from duties or customs."[39] On the 25th of July, the Treasury replied: "The Channel Islands are not excepted from the operation of the Customs Tariff Amendment Act of 1860 ... The Channel Islands are not subject to the same Customs regulations as the ports of the United Kingdom. It is therefore impossible with a due regard to the Imperial Revenue to extend to Jersey privileges against the abuse of which there would be no means of guarding."[40]

But worse was yet to come. In the same year, with Britain becoming increasingly concerned with the expansion of its world trade and the maintenance of the "Pax Britannica," a commercial treaty between France and England was negotiated, the first of a series of such treaties reducing French tariff barriers with neighbouring states.[41] Jersey was not included in the agreement, and Joshua LeBailly, now president of the Chamber, went to London in an attempt to clear up the situation, after the Chamber had been informed by Saint Malo authorities that "il devient difficile de vous demander de vos produits car les Îles Normandes sont exclus de traité de commerce entre la France et l'Angleterre, et notre douane est d'autant plus exigeante."[42] The Chamber protested that Jersey had "from time immemorial" been allowed to export to France its own goods as well as those of Britain, and pointed out that the spirit of the treaty was to make trade more free, not to restrict it. Following the well-trodden and previously successful path of claiming special status, the Chamber laid the case before the Secretary of State for the Home Department. It protested that the French interpretation of the agreement – as being only between England and France, and not therefore including Jersey, which they stated was a colony of Britain – was erroneous: "Jersey is neither a colony nor a conquest, but a peculiar and immediate dependency of the Crown, and in all commercial treaties the Channel Islands have always been admitted upon a footing of equality and as one with the people of the United Kingdom."[43]

Then, in 1863, the Jersey merchants in the Newfoundland fisheries also experienced difficulties with British legislation.[44] In what proved to be a last effort by the Chamber to influence the British government to assist the merchant cod fishery from the east side of the Atlantic, the Chamber wrote to the Secretary of State for the Colonies, pleading that the act recently passed by the Newfoundland government to levy duties on goods to Labrador not be given royal assent. They claimed that the act was injurious to Labrador and Jersey alike, that the people of Labrador were too poor to be able to pay for goods so taxed, that consequently the burden of the tax would fall upon the merchant, and that the fishery would therefore become unprofitable. In tones reminiscent of an earlier era, they pointed out that Britain would lose an important nursery for its seamen, and they professed outrage that Newfoundland should consider itself legally justified in passing legislation for Labrador when Labrador remained unrepresented in the Newfoundland House of Assembly. They condemned the Act as "illegal, arbitrary, inexpedient and impolitic."[45] But the days of the fishing ships were over, and Downing Street replied that the act was considered to be just:

The provisions of the Act are not unnecessarily oppressive. Consequently [the Act] ... has been left to its operation. At the same time ... if taxes are to be imposed by the Newfoundland Legislature on persons inhabiting the Labrador coast, those persons should be enabled to send representatives to the Newfoundland Assembly, and His Grace has accordingly recommended to the Governor that the Act regulating the representation should be so altered as to effect this object so far as it can be effected by the mere alternation of the law."[46]

Thereafter, the commerce with Labrador from Jersey was consistently weakened, until it effectively ceased in 1873 with the failure of the firm of DeQuetteville after the crash of the Mercantile Bank and the Joint Stock Bank in Jersey.[47] At the same time, the Chamber of Commerce became increasingly caught up in the growing general depression that was to reach epic proportions by the 1870s. After 1863, it concerned itself more and more with internal problems of the island, and its initial function as a watchdog over the essentially mercantilist economy of Jersey was rapidly eroded.

The principal role played by the Jersey Chamber of Commerce in the cod trade was to create connections into the core structures of the British Empire, thereby enabling the cod-trade merchants to have some impact on both the metropole and the fish markets abroad. However, it also achieved merchant solidarity in Jersey over such matters as labour disputes. In 1786, for example, the ships' carpenters of St Helier sought higher pay from their merchant employers. The records of the Chamber contain references to meetings between carpenters and merchants, along with an account of the merchant strategy devised to combat this labour threat:

It appearing ... that ships' carpenters insisted on an increase of their wages, it has been unanimously resolved that the request should not be granted, and that any Member found guilty of giving the said carpenters ... more than the wages they had therefore ... shall forfeit the sum of 300 livres French currency for every vessel on which workmen so paid shall have worked, which sum will go to the general stock. It is further resolved that, should they so persist, that workmen must be procured from England or elsewhere. Whatever expenses any Member may be liable to above the usual wages of those of this Island will be made good by the Chamber.[48]

The power structure of the Jersey commercial system is best regarded, then, as a nested hierarchy. At the bottom were the labourers and artisans of the island (and, by extension, of the Gulf) who operated within the mercantile structures of Jersey. As a collectivity, the merchants directed the commerce of the island through their Chamber. The Chamber operated as a hinge in island power: it had control of the commercial structure through the component firms, and it had influence with the higher political power structures of the States of Jersey, the lieutenant governor, and the King in Council. The States and the Chamber of Commerce were, in a sense, in contra-distinction to one another. The States represented Jersey itself as a domestic and agricultural, and also seigneurial, community. The Chamber represented Jersey as an island of merchants whose eyes were always fixed on far-off horizons of trade and commerce, and whose contribution to Jersey's economic well-being resulted from the good management of exogenous resources that created profits which were returned to Jersey. So long as landed interests were paramount in the island's economy, the States were the dominant force for the protection of its interests. With the rise of the cod trade, the Chamber was created and grew in power. By the 1840s, the States and the Chamber were combining interests to create the needed harbour facilities for an expanded trading (and therefore sea-based) economy. Ultimately, the merchant families of the Chamber came to hold office in the States and to blend these two functions: it was a misappropriation of combined business and States funds that brought about the collapse of the cod-trade empire's financial base in Jersey and created a simultaneous crisis in the States. Thereafter, merchant interests in the States decreased and the landed interests again began to rise. At the same time, the merchant structure of the Chamber of Commerce was altered as the fish merchants collapsed. The Chamber turned to the consideration of Jersey domestic products as the basis of future prosperity and, from that time on, it appears to have ceased to play the dominant role it had exercised in the sea-faring and trading days of the late eighteenth and nineteenth centuries. In its heyday, however, the performance of the Chamber was remarkable. The Jersey merchants achieved a unity of policy considerably greater than that won by comparable units of regional commerce such as those in New England or the West Country.[49] What the town of Poole could achieve,[50] the "nation" of Jersey achieved – and here lies the key to understanding the unusual integrity of its mercantile community. For Jersey was in some ways almost like a city state, being a "city" in terms of its small size and organizational coherence, and a "state" in terms of its political influence, which arose out of the ambiguous constitutional position of the island. This distinctive solidarity of interests provided invaluable stability at the metropolitan apex of the Jersey merchant triangle.

CHAPTER FOUR

The Merchant Triangle
in Action

It's not a game. Ye cursed blind fool. We gits ready for the fish year after year, that's all. And we waits. And out there, they knows we're waiting ... We took what we could get.[1]

Typicality, continuity, and change are all essential components of any study that covers a long period of time: what is sought in this chapter is the simultaneous portrayal of that which was standardized and that which was flexible or changing in the Jersey cod trade. Basic structure did not change once Charles Robin had put his finishing touches to it, but of course within that framework many things shifted about. Thus, markets in Brazil opened up, Caribbean contacts declined, the Naples market became pre-eminent. Likewise, new outports were established as time progressed, and agents became more and more important in the New World as the Robin family began to direct its business from Jersey rather than from the coast. Over time, too, Gaspé's population increased, Gaspé Bay became a free port, and there were also shifts in the wider international framework of the trade.

Nonetheless, it is possible to identify three stages in the business as reflected in the case study of CRC. The first was the early years of trial and error to about 1800. The last was the years after 1870, when the international economy began its downslide toward the Great Depression of 1873–96 as industrialization took hold of the North Atlantic economy at large, and the merchant triangle was put under increasing general stress. The middle years were the years of least stress and maximum expansion of the business, and it is therefore one of these – 1840 – that is used in this chapter as typical of the normal functioning of the production apex of the system. Mindful of the need to demonstrate both continuity and change, however, I shall also look at various aspects of the trade in terms of the fluidity that must be an essential component of any enterprise which survives over one hundred years. In the first part of the chapter, therefore, I examine the problems involved in commencing the fishing season at Jersey and in the Gulf and then describe the complexities of marketing, moving backwards and forwards across time as needed. Thereafter, the fishing season itself is detailed, using 1840 to present a cameo of this central, and more standardized, part of the trade and to show how its various elements fitted together from the perspective of the production apex of the trade. Finally, I discuss management of the trade as a fully functioning system, including the metropolitan organizational and financial components of the business as well as the familial structure of CRC itself. This chapter, then, is a picture of the merchant triangle in action.

PREPARATION FOR THE FISHING YEAR

Figure 10 shows the generalized operation of the system in a hypothetical typical year.[2] This is the system whose essential elements were put in place during Charles Robin's time in Paspébiac, to which markets and fishing stations were occasionally added during the century of Jersey Gulf operations, and in which slight shifts occurred from time to time in response to political, economic, or even climatic (such as a hard winter or a late spring) events.[3] The critical factor in the cod-trade year was timing – the always variable and sensitive point in the management of the fishery and the point at which luck and business acumen came together. This was equally true of operations in the Gulf (which were heavily dependent on weather and the biological and environmental requirements of the resource[4]), of market strategies (which had to contain effective timing for arrival at market), and of operations in Jersey (where the timing of the departure of the cod fleet for the fishery was crucial, and was linked to the seasonality of the Gulf fisheries).

The cod-trade year started in Jersey with a Spring Communion, held in the parish of St Brelade's for those parishioners who were to leave from St Aubin for the New World.[5] As early as February (see figure 10), the merchants of St Helier and St Aubin were starting to prepare their vessels for the annual exodus to Newfoundland and the Gulf, acting on instructions or advice from agents or part-owners resident at the cod-fishing establishments. Here, for example, is part

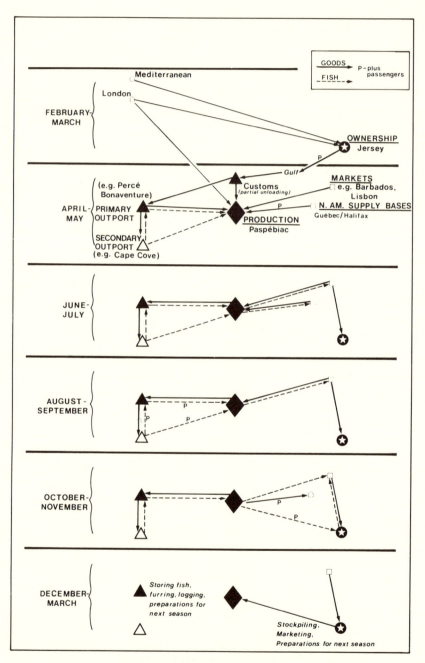

Figure 10. Generalized Operation of the Cod Trade

of Charles Robin's 1779 instructions to his Jersey headquarters for the following year's activities:

The "St. Lawrence" is intended to return here to take our fall fish ... "Truth", "J.F." and the new ship to take next year's returns home, the "Kingfisher" to be here in time to make a trip to Quebec after our provisions and be back in time to sail from here with a cargo of new fish by 25th July to run twice ... Herewith you will find the vessel's inventories and a sketch of what we may have remaining on hand. Nothing more can be done as our accounts cannot be settled before the Spring when the fall fish is received.[6]

The first order of business was the acquisition of stores from Jersey, London, and the Mediterranean,[7] if they had not been acquired already in market transactions the previous year in exchange for fish. In 1777 this was the case with the *Hope*, for example, which was to obtain in Oporto a cargo of wine for Guernsey and salt for Jersey in exchange for fish.[8] A voyage from Gaspé to Jersey, where passengers were discharged, and then on to the Mediterranean for salt was also common.[9] During the eighteenth century, stores were amassed in Jersey for loading, though London stores could be sent either direct or via Jersey,[10] but during the Napoleonic Wars, both Jersey and West Country merchants began to sail more frequently from Liverpool. This was a result of both the danger of capture in the Channel and the increasing number of products produced in the Midlands for the fisheries and shipped via Liverpool.[11] The vessels had to be fitted out for the voyage, the crews hired, the provisions for the fishing stations put on board, and the labour to man the fisheries recruited. As figure 11 shows, the usual time of departure was March or April,[12] depending on the part of the cod-fishing area for which the vessels were destined. The establishments at Arichat and on the south coast of Newfoundland, for example, being ice-free, were among the earliest of the fishing stations to be furnished each year. Labrador, in contrast, was the last, because its sub-arctic climatic conditions meant that it took a long time for the ice to clear and the water temperature to rise to the point where the fish began to run.

Captains of fishing ships signed an affidavit before clearing port:

Personally   appeared ___(Name)___ , Master of the _(ship's name)_ of Jersey, now bound for ___(Place)___ on a fishing voyage on the Banks, who did make oath that in pursuance of an Act of Parliament made in the 10th and 11th years of the reign of King William III, entitled "An Act to Encourage the Trade and Fishery at Newfoundland," he has _(Number)_

SHIPS ARRIVING AT:	APRIL	MAY	JUNE	JULY	AUGUST	SEPT	TOTAL
ARICHAT	■■■■ ■			■■			7
PASPEBIAC & GULF	■■■	■■■■ ■	■	■			10
RICHIBUCTO, QUEBEC, HALIFAX	■	■■		■			4
FORTEAU, LABRADOR		■	■■■		■		5
"TERRE NEUVE"		■■■	■■	■■	■	■	9
Total	9	11	6	6	2	1	

Figure 11. Seasonality in the Cod Trade, 1830

men and boys for that service, 3 whereof are fresh and green, and all his Majesty's subjects, 7 of which are to receive wages and the remainder upon his share ...[13]

This was in order to take advantage of benefits given to the British migratory fishery: such ships did not have to pay naval office fees, but instead paid only one fee on entering the fishery in the summer and one when they cleared out in the fall, thus saving expense on trading voyages round the coast, during which they had to enter and clear port a good deal.

Passengers for the inshore fishery included agents and clerks (to attend to management and bookkeeping); carpenters, blacksmiths, and other skilled workers (involved in the construction of the physical premises required by the fishery, or in shipbuilding); Jersey shore crews; and the fishermen themselves. Of these last, it was explained that "some men are employed solely in the catching of the codfish, and in bringing it ashore; others in the carrying it to the spot where one person is employed in cutting off the heads, and another in ripping and gutting them, and some in salting the fish. Others are engaged in the transporting of it to be dried, and, at the least symptom of dampness in the atmosphere, in storing it."[14] A Gaspé-bound ship leaving Jersey (or Liverpool) would carry with her such supplies as: oakum, food (Jersey biscuit, fruit, etc.), fishing tackle, sailcloth, soap, boots and shoes, wearing apparel, hardware, and, of course, salt.[15] Some of the vessels would carry passengers. Figure 12 shows the contents of one such vessel – the brig *Patruus*, master Moses Gibault – on her arrival at New Carlisle in May of 1844; the *C. Columbus*, which arrived there three days earlier, carried a similar cargo but no passengers. This is typical of the entire period of the 1780s to the 1880s.

At the beginning of the year, it was important to have all supplies in the Gulf for the commencement of the fishing season since, "if provisions come late, they are not converted into fish but into bad debts."[16] Salt was particularly vital, both for curing the firm's fish and for supplying the planters on the coast, and an insufficient quantity of salt in the area directly translated into a loss of cash – in 1776, for example, Robin calculated that he had lost £1,000 sterling through an inadequate salt supply.[17] The kind of salt used was also important, since small-grained salt provided a better cure;[18] the preferred variety was Lisbon salt, although in later years salt from Liverpool also became popular.[19] The quality of manpower – especially the Jersey *commis*, or clerks, who attended the stores and did the bookkeeping and the agents who supervised the various fishing stations – was, of course, crucial. In 1800, Charles Robin wrote to Philip Robin in Jersey that "if you don't furnish this Employ with a few good men, it must fall … Young Mauger cannot answer our purpose, he can hardly write and has never learned figures,"[20] adding, in an aside about the quality of Jersey supplies, that "such help and the bad shoes you send us are uniform …" Some of the shoremen also had to be extremely skilled: the quality of the cure depended on the aptitude of men such as splitters. CRC employed many French-Canadians on a seasonal basis for work such as this, and the Letterbooks contain yearly references to their engagement at the beginning of each season, especially those who were known to be good: "Si les saleurs Boulet, Nicolle et J. Bernaiche veulent retourner pour nous vous pouvez leurs accorder 1 ou 2 piastres par mois de plus que ce que j'ai spécifié et autant au bons trancheurs qui sont déjà venus pour nous et que nous connaissent bien …"[21]

In assembling supplies and labour for the fishery at the beginning of the season, speed was of the essence. The Gaspé fishery's main advantage lay in the earlier start of its fishing season compared to those farther north, especially that of the long-established Newfoundland fishery on the Avalon Peninsula, which was its major

Outport of Quebec,

New Carlisle.

No._____

Permit to *Charles Robin & Co* _____ to land

from on board the *Brig Patruus*, *Francis Gibault* _____ master

from *Jersey* _____ the following Packages having

been duly Entered, and H. M. Duties Paid or Secured thereon by Entry, No.

MARKS & NOS.	PACKAGES AND CONTENTS,
✓	Two Hundred Tons Salt
✓	Ten Cwt Oakum
✓	Twenty Pieces Lignum Vitæ
✓	Eight Casks Peas
✓	Six Quarter Casks Wine
✓	Three Hogsheads Vinegar
✓	Twenty Bales Fishing Tackle
✓	Five Bales Canvas
✓	Ten Bundles Sheet Iron
✓	Six Boiler Plates
✓	Twenty two Boxes Canada Plate
✓	Forty two Bags Biscuit
✓	Seven Bundles Steel
✓	One hundred & eight Stone Bottles
✓	Four Crates Earthenware
✓	1 Bell
✓	Four Boxes Candles
✓	Three Boxes Soap
✓	Twenty Six Suits Boat Sails
✓	Ninety five Coils Cordage
✓	Four Casks Boots & Shoes
✓	One Box and Seven Bales Slops & Woollens
✓	Two Casks Tinware
✓	One Sett Pintles Braces & Stirrups
✓	One Windlass Barrel and Pawls
✓	Two Hawse Pipes
✓	One Sett Steam Pipes & Deck Scuttles
✓	Cogs for a Winch &c
	Twenty and Fishermen & their Luggage

CUSTOM HOUSE, NEW CARLISLE,

16th May 1844

H Kavanagh

SUB COLLECTOR.

Figure 12. Contents of the Brig *Patruus*

competitor. Everything had to be ready when the cod started to run: the planters had to be supplied for the season, which meant that the credit system for that year had to be set in motion, and nets, gear, and goods had to have been supplied to the resident population. The ships which brought out the Jersey crews and agents were also those which took the potentially valuable "old" fish (the first fish of the year) into markets that would still be empty and hungry if they could get there first. So those vessels had to be rapidly prepared for their return voyage, since it was imperative that the Gaspé old fish cross the ocean ahead of fish from the Atlantic coastal fishing communities (which were closer to the markets by as much as a week's sailing time) if it were to prime the market and command a high price, regardless of the actual quality of the cure.[22] This was tricky to do since, on arrival in the Gulf (see figure 10), the cod ships had to touch at a customs port for clearance inwards (initially Percé, later New Carlisle as well), drop off the men and provisions at establishments there, and then proceed to Paspébiac.[23] Shallops took the salt and supplies to the smaller stations, where captains, crews, and fishermen had to be deployed for the season, and then brought back the old fish which had been stored over the winter, not having been cured in time for loading on the last ship sailing for market the previous fall. Markets for this old and fall fish had to be selected as it arrived at Paspébiac for trans-shipment from the shallops onto the ocean-going vessels. Obviously, speed was vital if Gaspé's natural advantage of an early start to the summer fishery was to be maximized and its natural disadvantage of additional distance from the first markets of the year minimized.

As the fishing season opened in the more northerly stations, they too were furnished with provisions and men, some of whom might have been hired in other places within the Gulf, such as the Magdalen Islands.[24] Figure 13 shows (for the year 1867) the details of this process for the Strait of Belle Isle, where several Jersey concerns had establishments. This figure and the two tables which follow (tables 2 and 3) are intended to supplement figure 10, providing details deliberately chosen from different time periods to demonstrate that the basic structure remained the same although, of course, relative numbers might fluctuate and new stations enter the picture. As figure 13 shows, some vessels came directly from Jersey with provisions and men, others (such as those of LeBoutillier Frères) entered through New Carlisle, reporting first to headquarters there and then proceeding to the Strait. One ship collected labour from the Magdalen Islands, and the last vessel (from Cadiz) was almost certainly coming from there with salt, although this information is

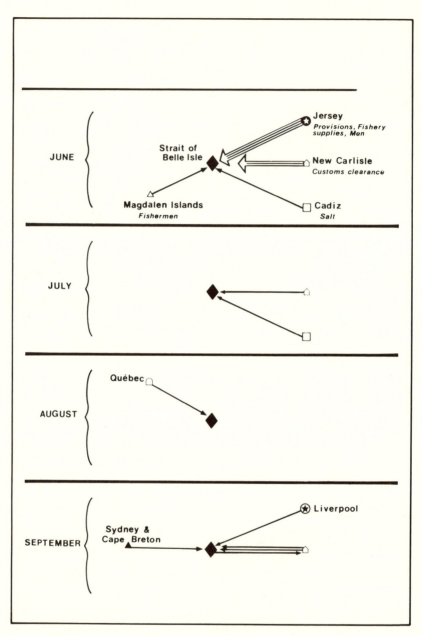

Figure 13. The Jersey Fishery in the Strait of Belle Isle, 1867

Table 2
Shipping Recorded Inward at New Carlisle From Main Cod Establishments in
the Gulf, May–November 1878

To New Carlisle, from:	78–150 tons	7–75 tons	Cargo
Caraquet	2 vessels	23 vessels	13,817
Percé	5 vessels	–	quintals, dry cod, cod
Arichat	3 vessels	2 vessels	oil, green cod, herring,
"Labrador"	1 vessel	–	oysters. All for lading
"River Nord"	1 vessel	–	and subsequent dispatch
Margaree	–	1 vessel	to market

Source: Fisheries Reports, 1879.

not given in the records. After goods and men were unloaded from
the vessels at Paspébiac and elsewhere, the fish from the previous
season were loaded and sent to suitable markets, where more supplies of salt and provisions were picked up. Fish from the outports
were also brought to the main establishments for loading for markets, and this collection and shipment of cargo from the various
stations continued throughout the season.[25] This part of the operation remained virtually unchanged from the early days on the coast,
apart from the addition of the daughter firm of LeBoutillier Frères,
which was established across the beach from CRC in 1838.

MARKETING

The establishment and maintenance of markets was a matter for
constant anxiety and consequent innovation throughout all the years
of the firm, since this was a vital part of the trade – the point in the
system at which the staple realized its value. But markets were by
their very nature uncertain, uncontrollable to a considerable degree,
and always open to competition from other British (as well as foreign)
fish producers.[26] Establishing one's product in the marketplace was,
as Charles Robin knew, largely a matter of good will and reputation;
in 1777, seeking markets among his Mediterranean connections, he
wrote to Thomson, Croft & Co., of Oporto: "It's all fine white fish
and I hope you'll be able to find us encouragements for further
connections ... I desire you will put off the cargo to our best advantage, either by selling on board or lodging just as you think best.
If this fish suits your markets as well as Lisbon we can every year
send you a cargo by this time, and we mean so to do."[27] By 1783,
he had established a market at Bilbao, through the agency of Messrs.

Table 3
Jersey Collecting Vessels Arriving at New Carlisle, 1875

Date	Ship	From	Cargo	Consignees
3 June	GDT	Magpie	12 qtls. dry cod	Le Boutillier Bros.
21 June	Hare	Caraquet	43 qtls. dry cod	CRC
8 August	Snowdrop	Percé	1,274 gallons cod oil	Le Boutillier Bros.
10 August	Northern Chief	Chéticamp	580 qtls. dry cod	CRC
21 August	Northern Chief	Margaree	660 qtls. dry cod	CRC
31 August	Northern Chief	Margaree	660 qtls. dry cod	CRC
13 Sept.	Northern Chief	Margaree	680 qtls. dry cod	CRC
16 Sept.	Ranger	Caraquet	3,026 qtls. dry cod	CRC
23 Sept.	Northern Chief	Margaree	660 qtls. dry cod	CRC
27 Sept.	Paspébiac	Arichat	450 qtls. dry cod	CRC
29 Sept.	Hare	Caraquet	200 qtls. dry cod	CRC
5 Oct.	Replevin	Caraquet	53 salt herring quarts	CRC
11 Oct.	Étoile du Matin	Caraquet	425 qtls. dry cod	CRC
11 Oct.	Union	Arichat	1,135 qtls. dry cod	CRC
13 Oct.	Beaver	Caraquet	460 qtls. dry cod	CRC
15 Oct.	Étoile du Matin	Caraquet	828 qtls. dry cod	CRC
16 Oct.	Paspébiac	Caraquet	786 qtls. dry cod	CRC
23 Oct.	Epopt	Caraquet	150 qtls. dry cod	CRC
26 Oct.	Adelina	Magpie	1,300 same, + 800 gals. oil	Le Boutillier Bros.
27 Oct.	Epopt	Caraquet	200 qtls. dry cod	CRC
27 Oct.	Replevin	Caraquet	180 qtls. dry cod	CRC
3 Nov.	Northern Chief	Arichat	1,020 qtls. dry cod	CRC
5 Nov.	Epopt	Caraquet	91 qtls. dry cod	CRC
5 Nov.	Replevin	Caraquet	20 quarts salt fish	CRC
5 Nov.	Fly	Caraquet	5,400 gallons oil	CRC
8 Nov.	Star of the Sea	Port Auxley	660 salt herring quarts	Le Boutillier Bros.
12 Nov.	Replevin	Caraquet	50 oyster quarts	CRC

Source: "Fisheries Reports" (Sessional Papers), 1876 for 1875.

Ventura Francisco Gomez & Berena, and was writing them that he would "flatter myself you will take us with you as old Friends," sending them 6,000 quintals every year.[28] Nor did he think only of the codfish side of the business, but (perhaps with an eye to diversification) wrote to William Gray of Boston: "My brother, Mr. Philip Robin, carries on Business in that Island [Jersey] and would be ready to serve any friends in America. Oak balks, Timber Planks and Staves to fill up would sell there very well, if you send vessels to the Baltic they might on their way land such a Cargo there for a trial."[29]

Table 4
Produce for Markets, 1795

Destination	Cargo
Jersey, for Spain or Portugal	3,826 qtls.
	3,038 qtls.
	850 qtls.
Quebec, for reshipment to London	52 large hhds,[1]
	15 tierces cod liver oil.
	1 poncheon skins and pelteries.
(Pelteries to be held in London because demand low)	3,350 qtls. for market of Jersey's choice
New York[2] (in the spring)	150 Bbls. salmon
	100 Bbls. herring

Source: Letter to CRC, 10/11/95.
Notes: [1] "worst of our inferior fish" because it sells well there.
[2] Proceeds from New York to go to London, because "we cannot import from United States."

In these early years of the firm, Robin was experimenting and would deal in any commodity that he thought he could market, provided it did no harm to the principal staple on which his trade was built.[30] Thus, in 1777, he wrote to Jersey: "If the Furr trade does not continue to hold out good, my intuition is to drop it as it gives a vast deal of trouble, and hurts the other business."[31] In 1778 he sent, on the *Bee*, the *Otter*, and the *Phoenix*, 6,300 quintals of fish, thirteen tons of oil, "and my furrs," which he calculated would fetch about £2,000 sterling.[32] Salmon were also part of this additional supportive trade, although initially Robin was reluctant to take it, since his vessels "don't go to proper markets."[33] By 1785, however, he was shipping it as a regular adjunct to the trade, and the practice of loading salmon as a supplementary cargo on vessels going to standard codfish markets in the Mediterranean continued throughout his career at CRC.[34] Timber was also on occasion a supplementary trade good, though in the early years it was more often used for shipbuilding at Paspébiac.[35] Later, however, it became an important adjunct to the trade, especially when the fishery was poor and adequate freight space was therefore available.

Typical exports for a season in the early years can be seen in the summary of trade that Robin sent to Jersey in 1795 (table 4); figure 14 shows the pattern of cod exports of the firm over the period 1838–70 (the years for which such data exist), and table 5 shows the fine detail of exports for 1864, to highlight another minor supplementary trade, the inclusion of the New York and Boston markets

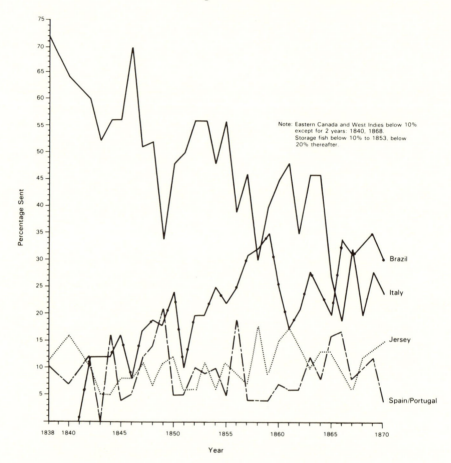

Figure 14. CRC Cod to Market, 1838–70 (MG 28 III 18, vols. 173, 175, 178, 181, 186, 189, 190)

for herring. The major points to be noted here are the geographical spread of markets, the strength of the Italian marketplace (in which Naples was dominant), and its relative decline once the Brazilian market was established. These are the outstanding features of the trade over the nineteenth century.

In the late eighteenth century, the Brazilian market was not directly accessible to firms operating in the British North American fisheries, since Portugal held a monopoly on that trade: all cod destined for Brazilian markets had to be exported to Portugal and then trans-shipped to and sold in Brazil by the Portuguese themselves. However, this also meant that the needs and preferences of

Table 5
Produce for Markets, 1864

Destination	Cargo
To Rio	3,000 barr. dry cod
	3,200 barr. dry cod
	? barr. dry cod
To New York	1,700 barr. herring (pickled)
	? herring (pickled)
	? herring (pickled)
Naples	6 "cargoes" cod
Oporto	large dry cod
Civita Vecchia	a cargo cod
Jersey	a cargo fish and cod
Bristol	60 tons cod oil
Total exported by CRC	41,107 qtls. cod
	126 tons cod oil
	2,200 barrels herring
Stored cod to be exported in Spring:	10,000 qtls.

Source: Fisheries Reports, 1865.

that market were known before direct access to it became available; in 1786, for example, Robin was sending the *Peace* to Oporto (with 1,182 qtls. fish) in August in order to be "in good time for your Brazil trade."[36] The Portuguese monopoly ended in 1807, when the British were admitted to a free trade with Brazil as a consequence of the flight of the Portuguese king to Brazil following Napoleon's invasion of Portugal. In 1812,[37] Newfoundland is reported to have sent its first direct cargo of codfish; by 1822, when Brazil declared independence, the market was established.[38] By 1830, CRC, now under James Robin, was shipping directly to Brazil, and by 1835 there was a Jersey agent in place there: Messrs. Thomas LeBreton & Co., Bahia.[39] In Naples, the other major market for CRC, the agent was Charles Maingay, the brother-in-law of Charles Robin's grand-nephew, Raulin Robin.

Markets were selected for a variety of reasons, some of them working in conjunction. Principally, as already shown, they were chosen according to the cure and quality of the fish being shipped and according to the state of the market (e.g., overstocked or clear, which affected the price being offered) or to the particular requirements of a given market. Climate, in both the fishing region and the marketplace, was important here. For example, the climatic conditions that required the late commencement of the Labrador fishery were a contributing factor to the difference in quality of its final product, the Labrador fish generally being of a softer, damper cure than that achieved in Gaspé, which had a longer number of sunshine hours.[40] Indeed, a variety of cures were effected, even within the Gaspé fishery, as a consequence of varying weather conditions, and so fish were graded according to their cure and then directed to the markets for which they were most suited.[41] The climate of the countries to which fish were exported was also instrumental. The hotter the country, the harder the cure had to be, in order to avoid spoilage. Very dry cures could best be effected with small fish. Consequently, the smallest fish, with the hardest cures, went to the hottest countries, and the larger fish of softer cure went to more temperate climates. Hence, Brazil received the smallest, hardest-cured fish, and those bound for the eastern Mediterranean were likewise of hard cure; moving westward toward the Straits of Gibraltar the cure became decreasingly hard and the fish larger. On the Portuguese coast, fish of relatively soft cure and large size were marketed. Over time, taste and tradition came to maintain this pattern: people quite simply preferred fish cured in the manner to which they had become accustomed because of past technological necessity, even when cooling systems, and therefore a variety of different cures, became available.

An additional attraction would accrue to a market that could offer a useful commodity in return for the fish: perhaps produce required for the fishery or for Jersey, or produce that could be taken elsewhere (that is, used as an entrance to the carrying trades). However, markets were never selected simply in accordance with whether or not a freight could be acquired at a market port after the fish were sold. The all-important factor to the merchant was the sale of his fish, and freights were a secondary consideration, sought to prevent a return in ballast if possible. Fish prices and the state of the market, of course, tended to go hand in hand – "I expect to be able to put nearly 3000 qtls in her [the *Hilton*] for Lisbon where I expect she might meet with a good market could she be dispatched no later than 12 proximo [June], for the high prices in the Bay of Biscay will attract more vessels there which I should imagine must occasion a scarcity of fish in Portugal"[42] – and these in turn were often combined with the need for a return cargo of goods required in the fishery. The letter quoted above, for example, continued, "By going there we ... [shall] be supplied with the best salt at a moderate price ..."

A glut in a market was to be feared as much as priming a market was to be desired. In 1777, Charles Robin instructed Jean Norman, captain of the *Hope*, to try to "*faire une prime*," something which had "not yet happened to us."[43] Gluts, on the other hand, could not always be avoided, and a good year in the fishery often meant that prices would drop as the market became saturated:

Ten more cargoes of baccalao have arrived ... Two of exceeding fine fish from Placentia, two-thirds large of this at 5 dollars ... five still remain unsold: the buyers being daily more terrified on so many arrivals [and] averse to engage in small fish ... We have advised both Captain DeCaen and Valpy to proceed to Italy, but they are alarmed at the quantity they say has gone forward to these markets ... We should be glad some of the others would proceed and by their lightening our market, for the present spare us the mortification of submitting to the low terms proposed by the buyers.[44]

Indeed, markets were spatially relatively concentrated, and it was a simple matter to proceed from one Mediterranean port (for example) to another if conditions of sale, or purchase of return goods, were poor in one location. Markets that could exchange fish for supplies required in the fishery (such as salt, which "il faut toujours en prendre"[45]) were not difficult to find, and those which provided produce for Jersey were always attractive – "We advised him [Captain DeCaen] to proceed with ... salt to Cadiz there to complete his cargo and make out bills of loading at said place for the whole. We shall load on him 4 hogsheads of our wine and perhaps the 20 pipes of brandies you have been pleased to order if can be had for our satisfaction"[46] – while a local writer in Jersey in 1837 observed that "the trade of this Island with Spain and Portugal consists in the sending there cargoes of codfish from our fisheries, and in the import here of wine, brandy, fruit and salt."[47] Again, the produce required for the fishery and for Jersey might be combined and sent to Jersey, that which the fishery needed then being sent out the following spring with the supply vessels. Typical of this would be the following: "*Dit-on* left Barbadoes 31st ulto [March] with 60 puncheons, 78 hogsheads 1 barrel molasses (10584 gallons), 54 barrels sugar at 3.60 (11140 lbs.) and 46 barrels sugar (9958 lbs.) at 3.50."[48]

Tables 6 and 7 show how this general pattern appeared from the Gulf perspective, with goods from Jersey and Bristol entering as "general," likewise goods from Barbados, which would comprise provisions such as rum and molasses, as well as salt, perhaps, from Turks Island.[49] Goods (such as fruit) landed at Jersey from market would be sold there and perhaps in Britain as well.[50] This selection of a port because it sold an exchange good (destined for neither Jersey nor the fishery but for sale elsewhere) which could be acquired for fish cargo was yet another variation in market strategy, and CRC

Table 6
Jersey Shipping at Percé, 1875 (Jersey Ships)

Ship	From	Cargo
Entered		
Hematope	Jersey	general
John Clarke	Jersey	general
Dawn	Cadiz	salt
Bolina	Figueria	salt
Heroic	Jersey	general
Dit-on	Jersey	general
Tickler	Bristol	general
Snowdrop	Cadiz	salt
Warrior	Barbadoes	general
Comalo	Barbadoes	general
Hematope	Barbadoes	general
Marie Georgina	Labrador	fish
Hebe	Liverpool	salt
Juventa	Runcorn	salt
Departed		
Dawn	Rio Janeiro	2,057 tin de pois
Heroic	Barbadoes	fish
Bolina	Naples	2,270 qtls. cod
John Clarke	Jersey	100 ton b. de c.
Dit-on	Italy	1,500 qtls. cod
Juventa	Rio Janeiro	2,152 tin. de pois
Zig Zag	"market"	fish

Source: 39 Vic. *Doc. de la Sess.* #5, A, 1876.

might actually purchase such a cargo itself, or freight it for another merchant. Either way, the produce of the fishery was often the means by which a CRC vessel could enter the potentially lucrative carrying trade once its fishery-related voyage was terminated, thereby earning extra returns in the off-season. In the case of the Brazil trade, for example, "there are ... vessels which annually proceed there with cargoes of codfish from the fishing stations, and return to Europe either with cargoes of produce on account of the owners, or on freight ... The return cargoes in Jersey vessels are usually sent to some of the continental merchants; sometimes a cargo is landed here to be transhipped [sic], and at others, for home consumption ..."[51] Nonetheless, such options were always considered in accordance with whether or not prices for cod in such ports were suitable; subsequent freights remained a secondary concern. Only when the best possible sale of codfish had been assured did the

Table 7
Summer Fishery – Market and Return Supplies, 1863
Gaspé Peninsula

Market (port)	Freight	Comments
Brazil (Rio)	3,200 barr. cod	Returned Oct. for next cargo.
Brazil (Rio)	3,200 barr. cod	Returned 15 September
Naples	3,800 qtls. cod	–
Civita Vecchia	3,000 qtls. cod	Returned October
(a) Québec	fruit	from Palermo then
(b) Jersey	flour	from Quebec; then
(c) Paspébiac	salt and goods	arrived October.
Barbados	624 barr./122 tons	cod, herring, salmon
Naples	2,400 qtls.	–
Bristol/Jersey	oil and herring	
Naples	cod	–
Jersey	wood	–
Naples	2,700 qtls. cod	–
Barbados	1,000 barrs. oil	Returned September,
Malaga	3,000 qtls. cod	Took fruit returned 20 Oct. with goods.
Cadiz	2,000 qtls. cod	1,702 qtls. as freight for merchant
Rio	1,747 barr. cod	Returned 23 Oct. with coffee.

Source: Fisheries Reports, 1864 for 1863.

merits of a port for the purchase of other goods become important. For example, in 1860 Jersey headquarters wrote to Paspébiac: "I find on inquiry that Trinidad would be a cheaper place for the purchase of molasses; Barbadoes being a place of call for ships, the price of produce is generally higher ...";[52] or again, in 1871: "My last accounts from Rio – *Century* was chartered 45/- for Orders – may go on the Continent between Havre and Hamburgh."[53] Finally, if freight or another cargo which would move the vessel into the carrying trades could not be had, there was always available as a last alternative the old non-lucrative standby of salt, perennially required in the fishery: "We hear of *Reaper*, and CRC, arrival ... shingles by the *Reaper* sold $3.87 $\frac{1}{2}$ and the oats $2.53 per 160 lbs. No freight so the vessels will return with salt ..."[54] Put together for any one year, the pattern of decisions resulting out of these various market alternatives, together with the supply shipping into the Gulf, created the cod-trade shipping patterns, or "geometries," which Head has referred to as a little-known element of the eighteenth-century Newfoundland economy.[55]

Over the years, market choices shifted somewhat in response to a variety of situations that demanded re-appraisal of traditional patterns. Marketing decisions were shared to a degree by the Jersey and Gulf ends of the business, with (in general) CRC at Paspébiac usually having more input on decisions about New World markets, while CRC headquarters in Jersey attended to European markets almost exclusively. The Brazilian market, however, was as much a Jersey as a Gaspé concern, since much of the carrying trade devolved back onto Jersey (see chapter 6). The West Indies trade, in contrast, with its supplies into the fishery (such as salt, rum, and molasses) was more Gulf-influenced. After Charles Robin left the Gaspé coast, market control in general evolved more and more into a function of the metropole, while production was overseen increasingly by trustworthy agents in the Gaspé fishing stations, and instructions for the year's activities were sent to them from Jersey.[56] Both ends of the trade, however, kept constant information flows in action in order to co-ordinate market strategies, production strategies, and supply lines. Intra-firm letters on market information often dealt with such production concerns as quality control: "Our neighbour's cargo via the *Griffin* sold 17/-, but owing to bad quality is reduced to 12/- ... This we hear by last advances from Rio ... we cannot be too careful in tubbing ... better have less and good, which pays better."[57] For example, one letter informed Raulin Robin in Naples that the cargo of the *Larch* had been of poor quality and sunburnt, according to the Oporto agents (Hunt, Roope, Teage & Co.): "They were selling cargoes of Fortune Bay winter cure to remit 14/- per qtl. but they could not entertain such flattering results of the *Larch's* cargo. We must consider ourselves extremely fortunate if it remits 12/- per qtl. to pay cost and freight."[58]

The most vital market information was contained in market circulars, which all the major companies used and which were sent to firms by agencies employed by them in the markets. They included general market information, freight rates, exchange values of currency, and the names of vessels that had sailed for foreign ports, along with information on the cargoes they were carrying. With such data on hand, plus the advice of their market agents, owners were kept informed of the well-being (or otherwise) of their ships and cargoes. The problem, of course, throughout the period up to the 1860s was distance, and hence time taken for information to reach a central decisionmaker in the firm, who had then to communicate such decisions to the people who would implement them. By that time, market conditions might well have changed.[59] Before the age of the telegraph (pre-1860s), then, the art of this end of the business lay in locating, at all strategic points in the trading system, agents capable of making such decisions on their own if necessary. Where possible, permanent agencies were established in the markets, both to handle the firm's business and to become a market business in their own right.

A multitude of different circumstances affected the deployment of vessels for market. Shifting political scenarios, the possibilities of a return freight, the need for specific imports (such as molasses or salt) in exchange for the fish, or the likelihood of a glut in the usual market port (which would lower prices for the cargo or even, in extreme cases, prevent a sale altogether) all had to be kept in mind. Captains often developed expertise about, and contacts in, particular individual ports, and would be sent preferentially to those places about which they were best informed. Correspondence between agents in market ports and headquarters in Jersey also helped to increase understanding of local market conditions, and market circulars were very useful here. Much of the skill required to manage the trade seems to have centred on the ability to make good judgments about markets, and even then sheer luck could make the difference between success and failure of a shipment.[60] A classic example of this last occurred in August of 1798, when Charles Robin reluctantly sent a cargo of unsuitable fish to Portugal, expecting poor returns. The cargo, inexplicably, sold well, although Robin himself acknowledged that "all she had was our worst fish." Of his good fortune, he commented dryly: "In the event of the *Hilton* it must be concluded that trade is a mere lottery, for with such a cargo as she had, to have destined her for Portugal would absolutely have appeared the greatest absurdity in the world."[61] But this was rare, and while fate might intervene from time to time, it was never counted upon. Sheer commercial acumen and entrepreneurial perspicacity, coupled with hard work and a rigorous organizational structure, were the keystones of Jersey success in the Gulf – as they were of all mercantile success at that time.

THE FISHING SEASON: 1840

As figure 10 shows, once the first vessels of the season had been dispatched to market from the Gulf, the fishery swung into full operation. The spring fishery was the vital one – "Our fertile and benevolent month of May, which used to produce us at Percee nearly one third of the fishing," as Charles Robin called it sadly in a year when it failed from bad weather – since it was this fishery that gave Gaspé an advantage over more northerly ones, and it therefore had to be got underway as quickly as possible.[62] The complete fishing season lasted until the weather broke, around early November (see figure 10), and the pattern of loading ships for market and receiving salt and produce for the establishments continued until then.

The year 1840 was not a particularly good one at Gaspé. Operations began on the 27th of April, with the arrival of the *C. Columbus* at Paspébiac "after landing Mr. Balleine and 40 men at Percé the preceding day"; the goods destined for Percé, however, had not been landed there, since the wind had been too fresh to unload the vessel. As a result, Mr de LaPerrelle, agent in Percé, had to send the customs papers up to Paspébiac, "as I suppose Mr. Balleine will have gone up to Grand River" to take over his station there. The Jersey headquarter's plans for the year's operations were expected on the *Fisherman*, and all was in readiness for the annual round: the agents were in position in Percé and Grand River, and the farmers had started to prepare the fields. This agricultural facet of the Gaspé fishery economy presented something of an organizational problem for the Jersey–Gaspé merchant. Seed-time and harvest were critical periods for both agriculture and the fishery, since it was toward the end of seed-time that the fishery was preparing for the catching season, and it was about harvest-time that the fishery was involved in final preparation for market. Agriculture, however, was vital to the maintenance of the local economy since, although it produced only enough food for local consumption, it nevertheless provided the means by which much of the resident population survived the winter months.

On the 4th of May, the *Seaflower* arrived, and the *Fisherman* followed at ten o'clock that evening; by the 5th, the men were being landed from the latter, and it was hoped to "have them all at work soon."[63] Preparations for collecting the old fish from the outports were underway, and James Robin in Jersey was informed that "Captain Smith will bring the fish from Bonaventure, which he is now ready to take in";[64] Creighton and Grassie, the firm's Halifax agents, were told that the *Columbus*, the *Fisherman*, and the *Old Tom* were all in port, and the *Old Tom* was soon to be dispatched with about 1,200 quintals of fish.[65] The *Old Tom* had arrived from Liverpool on the 5th of May, and the *Dit-on* arrived on the 19th, also from Liverpool, and were nearly ready to depart for Halifax: "The former has now 1000 qtls. fish on board and after two days of fine weather will be ready to proceed to your address with a full cargo ... The *Dit-on* is now discharging her inward cargo at Caraquet, and on her return will be sent to collect her load at the different outports, when she will also proceed to your address." The CRC arrived on the evening of the 19th "in 31 days from Liverpool,"[66] and all this activity in the area was reported in a letter to James Robin:

I have now to advise the arrival of the *Old Tom* from Percé on the 5th inst. with 905 qtls inferior fish ... Mr. de la Perrelle, having come up by *Old Tom*, sailed on the following day for Caraquet on board the *Dit-on*. I expect her back soon with about 100 qtls fish they have there, when she will be laid up to await the arrival of Captain Nicholas LeGros, it being high time that his brother should proceed to his station. Mr. George [LeGros] was landed at Percé so that he can attend on the room until the arrival of the Captain. I hope CRC, and *Patruus* will soon arrive [CRC arrived that evening] as their crews will be wanted. *Vincent* sailed yesterday for Percé and Grand River with the rest of the crews they had here, with the exception of a couple of hands who have gone on board *Dit-on* to assist in giving dispatch.[67]

It was now time to provision the stations and to arrange payment of commission agents:

The *Unity* sailed for Quebec on 14th inst [May]. By her return I had ordered from Mr. Symes 250 lbs of Biscuit, 20 cwts butter, 50 barrels pease, 6 kegs tobacco and 8 chests twangay tea, 2 ditto, congou ditto, 24 half boxes glass and the oak for the new brig. I gave him directions to draw for the shipment on Messrs. DeLisle & Co. forwarding them at the same time invoice and acct. current with directions to send it on to you. We are well stocked with flour and require none at present. When I have prices from Quebec and Halifax to compare I shall decide where to order from ...

On the 20th of May, the CRC arrived and made ready to unload. The *Old Tom* continued to load for market,[68] and the first signs of the fish run were noted: "The *Peace* returned yesterday [May 24] from Grand River and *Vincent* this morning from Percé. The latter reported that the Canadians arrived last evening, and that the day previous a boat which had gone had brought back 50 fish, that few capelin were caught."[69]

Problems, however, were in the offing. Local fishermen were being distracted from both the fields and the imminent fishery by two wrecks on the coast which the residents sought to plunder, thereby "neglecting both their farms and the fishery," to the detriment of the company and also – or so CRC argued – themselves.[70] Given a shortage of cash in the area,[71] and the distraction caused by the wrecks, headquarters in Paspébiac resolved that it would not supply "more than half the usual number, and those I leave off will, I find, bring their fish green in preference to going at LeBoutilliers."[72]

The firm, that is, would be protected from heavy debt being incurred through the credit system, without losing the custom of the fishermen (and therefore their fish) in the process.

The end of the first part of the seasonal round was now approaching. The *Patruus* and the *Lady Harvey* had arrived, the former was unrigged, and her captain was preparing to unload the goods and proceed with his crew to his station if the *Vincent* did not arrive in the meantime. The local fishermen, not yet finished planting their crops, were due to start the fishery, and construction of a new ship was well underway: "The carpenters are getting on well with the new schooner. They've finished planking outside and are now trimming her inside for sealing. Her beams are all in and kneed. The blacksmiths have chiefly been employed about her work." The *Diton* was almost ready to take salt and flour to Caraquet for the season and proceed to Grand River to take on board the old fish for sale in Halifax.[73]

By the end of the first week in June, the firm was preparing to ship fish to Halifax and to secure a return cargo of provisions. Across the barachois, LeBoutillier Frères were likewise loading their vessel, the *Teaser*, for market.[74] The fishery was also getting underway: "The last accts. from Grand River Mr. Balleine says our boats and ships' crews landed 204 qtls. and capelin plenty. They averaged 8 qtls each but [were] disturbed today [June 9] by strong currents."[75] With the assurance that there was plenty of capelin for bait for the summer fishery, CRC felt secure in the prospects of a catch, and headquarters now turned its attention to setting prices for the season's fish: "We are prepared to receive all the green fish that will come to us, at last year's rates, what we cured having left us a small gain. At Grand River where the fish is extremely small they will reduce their prices"[76] – that is, a small profit could be expected on green fish which CRC would cure itself. By this time, both CRC and LeBoutillier Frères were getting their ocean-going vessels ready to take the first cargoes of the year to Jersey or the Mediterranean. "The *Susan* arrived at Shippegan 1st inst [June]. *Teaser* arrived 4th inst. She is unloading and takes on fish and lumber to return to Jersey." Captain Balleine in the *Vincent* was also preparing to load for Jersey, unless James Robin instructed otherwise, and the CRC was taking 3,000 quintals of marketable fish on board for Naples.[77] A letter to James Robin dated the 16th of June gives a detailed picture of the summer fishery in full swing:

Judge Thompson has arrived after a long passage of 52 days. She delivered

470 hhds of salt at Percé and Grand River and arrived here yesterday with 832 qtls of fish. We are shipping on board today and will have her filled up tomorrow but she may be detained a day or two longer for repairs on her mainsail and topmast ... We have a gang at it to expedite it as quickly as possible ... Mr. Fauvel writes from Percé 11th inst. The fishery continues slack, the boats averaging 2 qtls daily with plenty of capelin. On that day they reckoned 28 qtls per boat being 30 qtls ahead of last year ... At Grand River on the 13th inst. their boats averaged 20 qtls and capelin in abundance. Here we have done much less. The best boat may have 12 qtls. Here there is plenty of bait but little fish, say one to two qtls per day. Mr. delaPerrelle says that the fishery commenced at Miscou the 8th inst. ... reporting fish and bait in abundance. The *Fisherman* are unloading and ready to take on. The *CRC* has yet 200 hhds of salt on board The *Seaflower* is discharged to 300, and the *Patruus* 410 ... The *Vincent* is ready with salt and goods to Grand River and Percé. The *Peace* will load with salt and flour for Grand River. According to news I may receive per mail today I intend to order 300 barr. flour from Halifax per *Dit-on* after which the balance remaining in the hands of Messrs. C. and G. will not be very great ... The new schooner is in a forward state. Hope she will be ready in a fortnight.[78]

A letter of the same day to Creighton and Grassie indicated that the *Judge Thompson* (now belonging to Isaac Hilgrove Gosset, a partner in Jersey who also did some small business on his own account through CRC) would be ready to sail for Civita Vecchia in "a couple of days" with a cargo of fish "on his account"; in the same letter was enclosed the bill of lading for the *Dit-on*, bound for Jersey.[79] The next day a letter was written for the Civita Vecchia agents dispatching the codfish bought from CRC by Gosset, and noting the early start of the Gaspé cod fisheries that year. The letter added that "we are in hopes of being able to dispatch our first ship, the CRC, for Naples in all [?] July. The *Fisherman* is to follow for your port and we hope will be dispatched the 10 to the 15 August and likely to be the first arrivals with new fish."[80]

The *Judge Thompson* sailed on the 27th of June. The *Victoria* had arrived from Cadiz and was discharging salt at Cape Cove. The *St George* arrived from Antwerp and, by the 29th of June, was headed for Cork.[81] A month later, the CRC should have been ready to sail for Naples, but fish was scarce,[82] and she was not finally dispatched until the 4th of August, with 4634.2 quintals of dried cod "which being of superior quality and suitable size will we trust arrive in time to prime your market and remit a good price."[83] In August, the plans were to have the *Fisherman*, which had now "commenced trading,"[84] ready in about eight to ten days for Civita Vecchia, the *Seaflower* was to follow for Ancona, and the *Patruus* was also to sail for Civita Vecchia some time in October. All was progressing normally.

By the 10th of August, the situation had changed: the fishery was poor, bait was scarce, and the shoremen were therefore to return to their homes at the end of the week. The crops were injured from lack of rain: there were only thirty to forty tons of hay taken, half of it cut, and three tons housed. The only positive sign seemed to be that late rains would improve the potatoes.[85] The 20th of August saw the *Fisherman* dispatched to Civita Vecchia in the hope that, although the season was advanced, she might arrive in time to prime the market.[86] The *Seaflower* was now loading for Ancona, and was expected to sail in ten to twelve days.[87] A letter of the 21st of August to I.H. Gosset showed him to be in no better fortune than the Company at Paspébiac: his ship, the *Broad Axe*, had arrived at Gaspé after being damaged by a mishap at Cadiz which had caused her to be delayed.[88]

At the beginning of September, headquarters was writing James Robin that both agriculture and the fishery had been neglected that year because of the wrecks in the spring, but that CRC had been so careful in the choice of which fishermen it had supplied that it was unlikely now to "contract debt":[89] its policy of caution had paid off, especially given the poor fishery that year. By the 13th of September, the fishery was beginning to wind down. The *Unity* was loading at Newport for Jersey, and was expected to sail in the course of the week. "The *Superb* which came over from Labrador was loading fish from Captain Vibert and Mr. Caen at Cape Cove for Jersey, to sail the beginning of the past week." The results of a poor fishery could be seen in the shipping arrangements as well: "You will have seen that the *Vincent* does not go home. Indeed I fear you will have trouble to make up cargoes for the other vessels, all the summer fish being nearly collected and the fall fishery being very slack owing chiefly to the scarcity of bait. I expect we will be obliged to ship more inferior to fill up the *C. Columbus*."[90]

By the 20th of September it was clear that the fall fishery had failed, and the Paspébiac agent wrote to James Robin with plans for winding up the year. The *Daemon*, he said, was to sail from Gaspé soon, and he had decided to fill up the *Dit-on* on her arrival from Caraquet, in about a week, with "about 700 qtls" of inferior fish and dispatch her for Bilbao. He planned to ship any surplus of marketable fish (though he doubted if such would exist) by the *C. Columbus*, which had 78 quintals of inferior fish on board along with "the oils from Percé and Grand River." The remainder of the shipping was to be disposed of as follows:

On board the *Patruus* 2440 1/2 small marketable, and on board the *Old Tom* 1553 large qtls. The *Peace* is at Port Daniel. *Nouvelles* and *Portage* will pick up about a half a load. The *Storm* and Grand River [the fishing station] will get about 300 qtls whereof 100 qtls large. The *Wasp* is at Bonaventure weighing and getting her bottom cleaned. She ought to get about 600 qtls. The *Vincent* is ready to proceed to Grand River for a cargo of inferior for the *Dit-on* or the *C. Columbus*. After the return of these craft we will have a sufficiency to fill up the *Old Tom* and will then dispatch her. Captain LeGros came up per *Vincent* a couple a days ago and has commenced rigging. The *Patruus* and the *C. Columbus* should be ready to sail about the first week of November.[91]

By the 1st of October, the extent of the failure of the fall fishery was such that "the *Argus* and the *Bellona* have had to load timber and Mr. Perchard is short of 200 qtls for *Guernsey Lily*." The *Unity* had returned from Quebec City with the provisions ordered earlier, which would keep the stations supplied during the ice-bound winter, and headquarters planned to keep enough supplies on hand to not have to import from Europe in the spring when the ice cleared out of the Bay, "as we get them cheaper and more suitable in Québec."[92] On the 2nd of October, Captain Nicholas LeGros was instructed to proceed to Jersey with the *Dit-on* (which was laden with a cargo of small, inferior fish) for orders as to where to market the cargo.[93] The *Old Tom* was to go to Oporto with 1,887.1 quintals of large, dry fish;[94] the *Patruus* was not yet ready to sail for Naples, but headquarters was inquiring of Creighton & Grassie in Halifax what size of shingles were suitable for its market, and what price they would fetch.[95] Late in October, the year's accounts were being drawn up and the Quebec agents issued a bill of exchange for the sum of £1,111–1–2 sterling at 9% premium: £1,345–12–4 currency. By the 6th of November, the *Patruus* was bound for Naples.[96] But the difficulties that had pervaded the entire year were not yet over, for in November Paspébiac had to report to James Robin that the CRC had been forced to resail for Ancona, having found a poor market at Naples – a fitting end to a bad year in the cod trade.[97] The final report of the fishing year was sent to James Robin on the 7th of November: the *Patruus* had sailed with 4460.1.0 quintals, and the *C. Columbus* had "1416.0.21 qtls small marketable and 1424.2.0

inferior whereof about 320 qtls of haddock but which are mixed with the cargo," as well as oil and blubber. She also carried juniper tree-nails. Forty persons were on board her, including three passengers "whose passage money is secured here."[98] Thereafter, the establishments settled in for the winter, securing pork supplies and potatoes for the spring, when the whole routine of the fishing season would start again.[99]

THE INTEGRATED SYSTEM – MANAGEMENT, FINANCE, AND ORGANIZATION

Behind the sometimes frenetic activity that was involved in the operation of a commodity trade of this kind lay the hidden machinery of information flow, finance flow, management organization, and decision making that made the system work. The smooth functioning of complex networks of supply, production, and trading was made possible by an equally complex and highly organized business structure which supported and maintained the enterprises in the Gulf and Jersey during the course of the seasonal round. This structure ensured an efficient information flow between Jersey, the suppliers, the Gulf, and the markets, and permitted decisions to be made in the wider context necessary to the efficient functioning of the whole system. On those occasions when the information structures malfunctioned, complaints were immediate – especially in the case of Charles Robin, "overhasty"[100] as he confessed himself to be: "If it's your plan that the J.F. should land here you ought to have explained it for my satisfaction for I ought to be fully instructed in time of the business you mean to carry on as otherwise I am perplexed and must go through all by guess ... You don't even mention whether the *Friendship* arrived safely, nor how we are to have our Jersey goods. I hope you have taken some measures to secure me bread in case of a catastrophe. It would be a great hardship to find myself a beggar after my long toils."[101]

But continued information flow was not a problem just for a young business still finding its feet. It was a constant concern of the company throughout a century of business practice, and it involved all aspects of the trade, including the supplying of local planters – "Proper salt measures are at Percé to be used when the vessels arrive, which I hope will be earlier in next spring than the last, because until then I will be in the dark how to act in giving out supplies"[102] – and checking out market conditions and political situations, as in 1860, when it was noted that "there had been a reduction of duty of about 2/6 per qtl. at Naples which they seem to say will greatly increase the consumption there though on the other hand the disturbances in Sicily and which may reach Naples make it very hazardous to send any property there, and may impede the sale; according to the circumstances we may be obliged to order some of the vessels here for orders"[103] – and checking out market prices and freight rates, and making the resultant plans for a voyage: "The *C. Columbus* arrived at Rio de Janeiro 25th February and sold codfish 15$000 and Haddock 13$100 at 6 months from 11th March. She was on lay days, 45 allowed to load and unload a freight; if at Liverpool 50/- per ton, Cowes or U.K. 55/-. She will come here from the place she unloads in Ballast except she went to Liverpool when she would bring salt"[104] – and communicating price strategy for Gaspé for a season: "You will pay fish the same as last year's in barters, but for cash you will have to keep up the price and not have Fauvel and LeBoutillier to consult together to fix our price. We foresee that we shall have to pay 13 or 14/- at least. You will have to secure a first cargo per *Telegraph* with a dispatch as last year. For that you will have to secure the first handy fish for cash. There being a demand for Haddock here you may pay them 10/-."[105]

These isolated examples of the kinds of information flow that were maintained between the various representatives of the company in different geographical locations show only a very small part of the decision-making and information-giving structure of the firm. The art of the business lay, in considerable part, in managing a dynamic trade and production structure, which meant that information had to be as complete as possible for all areas, and that control likewise had to be as complete as possible in all areas. In other words, the firm had to achieve a level of organization that would allow it to respond to the flux of trade and production (which was essentially non-rigorous) in a flexible but organized manner. To remain inflexible in the face of altering circumstances – changing markets, bad years in the fishery, alteration in the political *status quo* (either in the Gulf, or in Jersey, or in markets) – was to invite disaster; but to respond to every circumstance in an unco-ordinated manner was to invite chaos – hence the constant care taken by the firm and its representatives to achieve maximum information flow. The commercial genius of CRC lay in its capacity to maintain overall unity of organization and purpose in its business dealings, while creating a controlling structure that was at once both rigorous and flexible.

The nerve centre of the Jersey merchant system resided in Jersey itself, the organizational and ownership base for the trade. Here were written the political petitions that sought to create and maintain a favourable politico-economic framework within which the trade could operate at maximum efficiency; here also was where many of the supplies for the fishery were assembled, where the exchange goods bought with the produce of the fisheries were filtered back into the system as supplies, and where other goods were purchased or manufactured for dispatch to the Gulf. Here skilled labour was recruited in the form of shore crews, clerks, and agents, and here also some of the unskilled labour was hired. Here the ships' captains and crews were contracted, and here the vessels returned at the end of the season. From here was deployed the capital that created the trade, and back here came the profits of the trade either directly, or indirectly through London.[106] This was home, the metropole, masterminding the workings of its outposts across the Atlantic, and the marketing of their produce, to generate commercial success for itself.

How the finances of this far-flung commodity trade were actually handled is extraordinarily difficult to research, given the serious lack of documentation on the metropolitan end of the system. The Gaspé ledgers of CRC show how the production apex of the trade worked (chapter 5), but only one metropolitan account book has been discovered, and it belongs not to CRC, but to a small company, Perrée Bros., a shipper operating on the edges of the Robin system in Gaspé.[107] As such it is very useful because the relative simplicity of the Perrée business, the manageable scale of its operations, and the fact that it was involved in the transportation part of the trade all combine to provide a straightforward analysis of the circulation of finance between the three apexes of the merchant triangle. Moreover, the accounting system the firm employed seems to have been roughly the same as that used by other Jersey companies of the time, as far as can be judged from the CRC letterbooks. What the Perrée account book cannot illuminate, of course, is the colonial landward aspect of the cod trade, because the firm does not seem to have been involved in that to any great degree (its account for 1845 shows that it spent very little on supplies for the Gaspé coast), but fortunately this is not a significant loss, since the CRC ledgers provide a wealth of detail on that part of the business.

Figure 15, then, shows the financial flows of this Jersey firm during the years 1847–1856. In the case of a small firm like the Perrée business (a shipowner who fished every season, outfitting himself and crew, but who did not become involved to any great degree in the supply aspect of a merchant fishery), provision and supply costs could be minimal.[108] On June 22, 1845, for example, Francis Perrée signed the following customs declaration:

The following goods all British and Jersey produce:

Goods	Provincial
10 bbls. flour ... value £10	5/-
2 bbls. pork ... £3-2-8 per vat	7/2
10 bags biscuit ... £5-10-0	11/-
5 cwt. oakum ... £5	- (no tax)
30 cwt. cast iron £6-14-0	- (no tax)
3 bags, 1 1/2 cwt. (?) (illegible) £3-0-0	- (no tax)
6 ? cwt. arrowroot £15-0-0	1-9-9
	2-12-11
	1/5 10- 1
	1/12 -11
	Total: £3-4-5

3 pkgs fishing tackle 30-0-0	
1 cask 28 ? for fishermen 22-8-0	
24 boat rods, 2 chains, 4 doz blocks	£49-0-0
for boats employed in fishery as	£ 4-0-0
fishing utensils and instruments	£2-10-0

That is, a total of £107–18–0 was spent on non-taxable goods, plus £448–6–8 on taxable goods, plus £3–4–5 tax: a total of £159–9–1 expenditure on the Gaspé establishments of the firm that year in imported produce.[109]

Figure 15 shows how many of the "banking" and financial functions of the firm were often handled from London,[110] in the manner described in chapter 3 for the firm of Fiott and DeGruchy, with owners receiving most of their profits from London at the close of accounts each December. In the case of the Perrée firm, however, the owners handled their own ship accounts,[111] since they were both the captains and the owners of their vessels. The finances of the Gaspé establishments, consisting of wages, store goods, insurance, and salt,[112] they left to London. They took their profits directly from the ship accounts, leaving a balance to be carried forward for "Ship Expenditures" at the beginning of the following season. The London agents[113] paid out to the Quebec agents, Messrs. Le-Mesurier, Tilstone & Co.;[114] to the Gaspé establishments; to Dean & Mills, of Liverpool,[115] for salt in April of 1848; to foreign agents such as Vito Terni & Co. of Naples, or Maingay Robin, also of Naples; and to Janvrin, Durell & Co., which was the Commercial Bank of Jersey. They received money from freight carried for various businesses, including foreign accounts, and although the owners received some money directly from ship-freight profits, the large sums

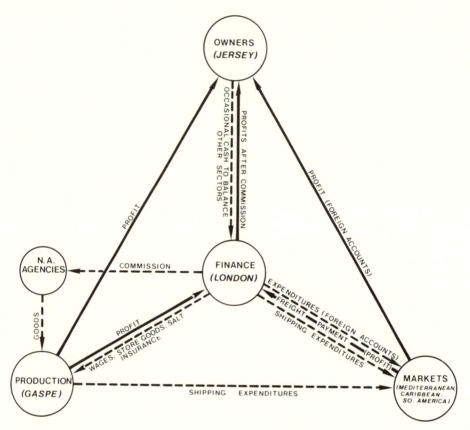

Figure 15. Financial Flows at Perrée Brothers, 1847–56

were handled through the London agents: for example, on January 7, 1848, £1,000 was received from Messrs. Dobree, Maingay and Company (later Robin, Maingay) for cargoes and freight. Profits from Gaspé were also paid to London, for handling. The Quebec agents did very little other than to invoice goods for the Gaspé establishments, for which they received a 2 ½% commission on drafts paid through London: a total turnover of only £313–4–6 currency occurred for Quebec in 1854 (that is, £261–0–5 sterling). London itself was paid at ½% commission, but on a much larger turnover. Balance remaining was then transferred to the owners – £1,390–19–10 in 1848, leaving a small sum which was paid into folio 48 of the ledger (no longer extant), presumably to cover operating expenses to the beginning of the next season.[116] The Perrée Account Book, unfortunately, is incomplete, and supporting documents such as the ledger have been lost. Therefore, while it does give an idea of relative cash flows in the firm, much of the detail is missing. The general pattern is, however, confirmed by both the Perrée letters and the CRC letterbooks, the latter providing some additional detail on capital flow from the point of view of the production and market aspects of the business.

Capital flow in CRC can be seen to have developed, like the firm as a whole, from initial structures set up by Charles Robin. In 1777, working under the system used by Robin, Pipon & Co., he engaged as London commission agents the firm of DeGruchy and Fiott. "Messrs. Hake of Lisbon remit us on the reception of the cargo, one third, and I hope you will do the same and through the hands of Messrs. DeGruchy & Fiott of London."[117] After 1783, as CRC, the firm continued to use Fiott DeGruchy until 1797, when John Fiott died and bankruptcy proceedings were started against his firm. CRC then switched to the house of Paul & Haviland LeMesurier in London.[118]

In markets, agencies – such as Messrs Hake in Lisbon and Messrs Thomson Croft in Oporto – were needed to see to sales of fish, acquisition of cargoes in exchange or for cash, securing of freights, notification of Jersey and London interests of the state of the market, and the financial transactions involved, such as payment of a ship's captain for any dealings that he had beyond those of the firm, or for commission. These transactions, both debit and credit, were then placed to the account of the cod-trade firm and drawn on through the London commission agent. Captain Thomas LeMesurier, for example, was paid thus:

Paspébiac 18 October 1783
At eight days sight of this my first of exchange (second not paid) Pay to the order of Mr. Thomas LeMesurier sixty eight pounds sixteen shillings sterling value received of him which place to acct. of Messrs. Ch. Robin & Co. as per advice from

Charles Robin

To Messrs Ventura Francisco Gomez and Barena, Bilbao.
Sterling No. 1 Ex° for £ 68-16-0 Stg.[119]

More details of the process can be seen in the Fiott Papers:

Enclosed ... sales of 661 qtls ... of Captain LeCheminant
N. Pd. £1760
ditto of Qtls 3000, Captain DeCaen
N. Pd. £10,904

To yr. credit in a/c current without our prejudice till in cash £1264 (11/1/1772)
You have adjoined invoice of ... sale shipped per Captain DeCaen amounting to £473-10-0. His receipt for disbursements pd. him £658-11-6 which we pass to yr. debit in a/c current £1132-1-6 ... (18/1/1772)
We debit you £3 3/4 each [for boxes of fruit], £15 which please to credit us in conformity ... (21/1/1772).

That is, there was complementary bookkeeping between the firms.[120] In like manner, accounts current for Quebec agents were sent back to London in the fall,[121] an account which, by 1794, was becoming a substantial sum for CRC – to the amount of £1,200 on average for "produce from Québec."[122] New York agents usually worked with London bills of exchange, either through the Quebec agents[123] or direct to London, "for we can import nothing from the States."[124]

Some cash transactions were involved with European markets, but the risk of privateering in the early years of the firm, as the Napoleonic Wars threatened, made the movement of specie hazardous, and the English expedient of transferring specie by warship to London was clearly not a solution for the island-based Jerseymen. Later, the system was based on bills of exchange. In 1791, for example, Robin wrote to his Santander agent that "silver dollars are the best and quadruples next in rank, golden dollars are the worst of all,"[125] and by 1792 he was warning Philip Robin to "take care not to get into any scrapes for the specie, or with the buyers with whom you must have a very clear understanding, and plain bargain in writing."[126] He himself adjusted financial strategy during the Napoleonic Wars so as to avoid risk: in 1797 he wrote to Philip Robin, "Considering the precarious state of affairs in Europe, I have given directions for the net proceeds of our vessel cargoes to be lodged in the States until further orders, to be placed in the stocks or otherwise advantageously placed till your further direction and I pray you will not hazard to remove it until its clear there is not risk at all, as it might so happen that we might be glad to find it there in a day of need."[127]

Charles Robin laid a sound financial basis for the firm, and by 1800, one-and-a-half years before his retirement, he estimated the value of the goods in Canada alone (salt, timber, crafts, and dry goods), premises not included, at upwards of £2,000 sterling.[128] In later years, the company continued to work on the same financial basis as that which he had laid down, working through Halifax, Quebec, and occasionally New York agents, and agents in European and Brazilian markets,[129] but always with most of the financial control being exercised from Jersey. The major areas of control which the production region in the Gulf administered were the organization of British North American supplies (such as at Quebec and Halifax), as just shown, and a variety of local activities directly involved in the production of the staple, such as the hiring of shoremen and the running of the seasonal round.

THE INTEGRATED FIRM — THE STRUCTURE OF CRC

So much for the organization of the trade itself. What about the dominant firm within that trade? In the eighteenth and nineteenth centuries, the horizontal integration of a firm was often achieved through the use of family members,[130] located at important nodal points in the commercial system, as a way of ensuring that loyalty and self-interest would operate together to keep the business profitable. This was particularly needed, in an age of poor communications, with a business that was geographically dispersed. As with so much else, the basis of such a *modus operandi* for CRC was laid by Charles Robin. In November of 1797, five years prior to his retirement, Robin seems to have set about putting the finishing touches to his long-term plans for CRC with the suggestion that his nephew John should be established as a correspondent for the firm at Liverpool to deal with transactions involving (at that time) the sale of ships and the purchase of salt. At the same time, he recommended that Philip and James, his other nephews, buy into the firm, each purchasing one-twelfth of the interest previously held by J. Fiott & Co., which was now bankrupt. "No better agents can be found," he said, "and they are to be preserved." Moreover, he proposed further capital investment in the concern *by its owners*, perhaps the surest sign of all that he was finally convinced that he had created a business that would survive and flourish: "I recommend to the Employ to find within itself whatever money will be necessary to carry on the business by launching out something if it is found necessary. In my opinion, it is a prudent measure and I recommend it."[131] In 1800, he pursued the issue further, writing to Philip Robin that "its [sic] absolutely necessary one of your sons should immediately be with you in the first place to relieve and assist you, and in the second to learn without loss of time the nature of the business in general" and suggesting that the Lisbon house of Axtell and Robin that John would be leaving should be replaced by a new house under "Mr. Leigh and James Pipon," the latter also related to the Robins through Philip's wife (see chapter 3)[132] and the former a manager in Newman, Land & Co. In 1802 Charles Robin retired, noting, "After my departure the business will be carried on by my nephews Philip and James Robin under the firm of Charles Robin & Co."[133] In 1815, the firm of Robin & King of Liverpool was established under John Robin and, in the ensuing years, the policy of horizontal integration of the firm continued until supply, production, market, and finance structures were all controlled by immediate family or kin.

PRC (Philip Robin and Company) was the "family firm" of the complex of Robin companies, and was almost entirely Robin-owned (see figure 16). CRC was the larger firm, and had directors in London and Liverpool as well as in Jersey and Gaspé. Frederick DeLisle, of the London firm of DeLisle Janvrin DeLisle, was one director, while the Liverpool director was John Robin, of Robin and King; Frederick Janvrin was also a director, as was Thomas Pipon of Surrey. John Robin for a while was a partner in the Lisbon marketing agents of the firm, Axtell and Robin;[134] in the 1840s C.W. Robin was in partnership with Isaac Hilgrove Gosset in shipping concerns involved in the South American trade;[135] and Raulin Robin was a partner in and later took over the Naples marketing agency Charles Maingay and Sons.[136] C.W. Robin's sons took over the old Janvrin-Durrell bank as Robin Frères, and this bank survived the 1886 bank crash that destroyed CRC itself. The firm of P. and F. Janvrin was linked to CRC through the London agents and through the shared directorship of Frederick Janvrin. Likewise, prior to Robin Frères, there were Robin links through C.W. Robin to Janvrin Durell & Cie. The wife of C.J. DeQuetteville was a partner in the firm of Nicolle, and C.J. DeQuetteville himself was attorney-at-law to Isaac Gosset. The original vision of Charles Robin, then, laid the basis not only for an efficient fishing firm in a newly settled region, but also for an effective staple trade operating in an international arena. Figure 16 is in effect a summary of his management strategy, based on family and kin, as it evolved to control an international commodity-trade venture at all three apexes of the merchant triangle.[137]

The preceding chapters of this book laid the empirical and structural groundwork for a comprehensive understanding of that merchant triangle. In this chapter I have looked at the integrated operations of the triangular system, where it had to be flexible and why, as well as what was constant in it and why. The result was a trading system that was spatially extensive, commercially complex and sophisticated, and clearly both stable and successful. Alfred D. Chandler, in his monumental work on the rise of the modern Amer-

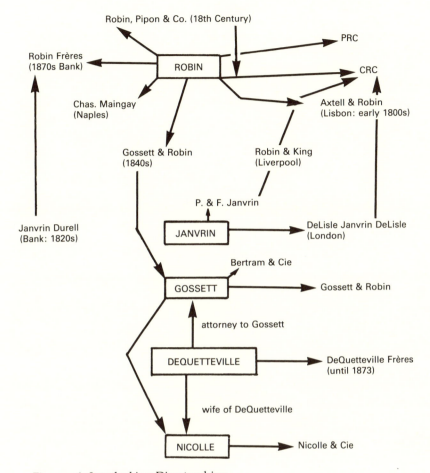

Figure 16. Interlocking Directorships

ican business, does not deal with the "traditional" (as he calls them) forerunners of the integrated firm in much detail.[138] In two chapters on "the traditional processes of production and distribution," he acknowledges the role of the "general merchant" as "exporter, wholesaler, importer, retailer, shipowner, banker, and insurer"[139] – a role that has been analyzed here in all its complexity – but, by concentrating on the United States, Chandler misses the functional sophistication and genuinely international structure of such enterprises. He has portrayed them in generalities, and seems to assume that they were in some way primitive. In particular, he failed to identify, let alone explore, the whole colonial-metropolitan relationship that was a central component and major motivating factor in this kind of commerce. It is to precisely this relationship that I turn in the next two chapters, to consider the question that arises out of this analysis of the merchant triangle: what was the impact of the cod trade on the two regions where it was of paramount importance – what did this staple trade mean to Gaspé and to Jersey?

PART TWO

Consequences of the Trade

CHAPTER FIVE

Consequences of the Trade: Gaspé

The fishermen are the main staff and support of the commerce of their country ... But they themselves, although they compose a large proportion of the population, are literally in bondage ... Their unprofitable callings have rendered them destitute of the means of improvement, and doomed them to perpetual servitude.   *Acadian Recorder*, 24 March 1827[1]

The economic history and historical geography of Gaspé remain to be written. There are, however, several useful sources for selected aspects of the history of the coast, of which Lee and Samson are the most recent.[2] Beyond this, the writings of Cooney, Mountain, Ferland, Ouellet, Blanchard, Langelier, and Chambers contain the best historical information on the region to date,[3] but none purport to be a definitive history of the peninsula. Nor is such a history to be found here: the purpose of this chapter is to assess the structural impact of the Jersey fish-merchant triangle on the potential for economic development that existed on the Gaspé coast during the period of CRC's residence there.

Generally, the settlement of an area will, *ceteris paribus*, in time generate a local economy: the establishment of a population is also the establishment of consumers whose demands are initially met perhaps by themselves (subsistence or home production) in combination with imports, but which can in time be serviced by a combination of imported goods and locally produced (i.e., manufactured) goods substituted for previous imports. One definition of the development of an area, then, is the eventual generation of a local economy based on domestic industry producing for a domestic market. Staple theory, stated in its more recent guise of the export-base theory of regional economic development, suggests that in newly settled regions such economic growth occurs initially around the export-staple sector as a result of linkage developments which arise naturally out of the staple base; subsequent diversification beyond that base depends on the success of the initial export sector. When a successful staple economy takes root in an area, capital accumulates there, incomes begin to rise, and the area develops a network of transportation links as the expanding export sector begins to create a hinterland and the economy generates an increasing demand for the circulation of goods, services, and people. Ultimately, a diversified economy emerges, fuelled by domestic production for the local marketplace; the old export sector ceases to be its leading edge and usually fades away.[4] Thus Watkins and Gilmour explained the growth of southern Ontario – but they also recognized that not all staples seem to have been equally successful in generating regional development, and a mythology has grown up in which some staples are seen as successful and others as failures, fish being denigrated as the staple with the poorest development track record of all.[5] The proposition, however, has not been so much tested as merely affirmed, and in this chapter, by examining the impact of the cod-merchant triangle on Gaspé, I also seek to discover if fish *per se* was responsible for the lack of development of the region or whether other factors were instrumental too.

Table 8 shows the production of cod by CRC as shown in the Paspébiac ledgers for those years for which data are available. Since other CRC stations, such as Arichat, were not included in the Paspébiac grand total, the table shows only the total production for the Gaspé area, which therefore underrepresents the total production of CRC. Even so, this level of output of cod makes CRC one of the largest fishing firms in British North America – it would have ranked twelfth among the St John's merchants in 1855 and sixth among all Newfoundland exporters in 1865. The comparison with Newfoundland is valuable because by 1874, when regional data first become comparable, Newfoundland produced 49% by value of all the fisheries (not only cod) in the entire Atlantic region. Clearly, the staple trade itself was highly successful throughout these years.[6]

Table 9 shows the 1831 population structure for the two counties comprising the Gaspé region: Gaspé County (along the south shore of the St Lawrence from Cap Chat to Newport) and Bonaventure County (on the north side of the Baie des Chaleurs from Point Mackerel to Restigouche). The population totals for 1825 have also been included to highlight the increase taking place in the populations of both counties. From the table, it was calculated that 4,006 (or 38.8%) of the 10,312 persons enumerated were under the age of fourteen years in 1831. The married population numbered 3,214 persons,

Table 8
Production of Cod, CRC, 1826–1877

Year	Quintals	Year	Quintals
1826	25,626	1852	31,654
1827	23,742	1853	32,481
1828	20,444	1854	29,296
1829	19,669	1855	32,391
1830	18,273	1856	33,473
1831	20,000 (est.)	1857	33,578
1832	22,000 (est.)	1858	34,304
1833	23,213	1859	42,806
1834	25,529	1860	42,495
1835	28,887	1861	43,570
1836	36,286	1862	49,230
1837	29,145	1863	53,071
1838	25,935	1864	44,061
1839	26,284	1865	51,241
1840	26,896	1866	56,587
1841	24,599	1867	45,144
1842	32,841	1868	48,848
1843	27,484	1869	40,874
1844	29,938	1870	47,157
1845	34,935	1871	54,571
1846	34,105	1872	54,702
1847	37,774	1873	52,242
1848	35,198	1874	46,586
1849	37,674	1875	41,365
1850	32,312	1876	50,848
1851	29,733	1877	53,220

Source: Robin, Jones and Whitman Papers, NA, MG 28 III,
vols. 174–5, 178, 181, 183, 186, 189, 190, 191–2. The figures
are taken from the summaries of total firm production as
recorded at the beginning of each year.

about the same as the number of single persons over fourteen years
of age.[7] That is, a superficial assessment of the figures would suggest
that the basis of a growing permanent population based on natural
increase had been laid: 70% of the population comprised married
adults and children, and 55.5% of this group were children under
14 years of age.[8]

It is, then, beyond doubt that both the population of the Gaspé
coast and the area's staple base grew in the years of Jersey merchant
enterprise in the region. How the Jersey cod trade affected Gaspé's
development is harder to establish. Certainly by the 1830s, after more
than sixty years of Jersey commercial enterprise on the coast and at
a time when the cod fishery was about to enter its period of maxi-

Table 9
Population Base of Gaspé, 1831

	Population 1825	Population 1831	Number of married males over 14 years	Number of married females over 14 years	Number of single males over 14 years	Number of single females over 14 years	Number of houses	Number of acres of occupied land	Number of acres of improved land	Immigration 1825–1831
Gaspé County (Newport to Cap Chat) and including Magdalen Islands	2,108	5,003	830	627	1,382	187	865	37,850	6,597	3
Bonaventure Co. (Restigouche to Pt. Mackerel)	4,317	5,309	638	1,119	1,232	291	939	98,364	12,090	112
Totals	6,425	10,312	1,468	1,746	2,614	478	1,804	136,214	18,687	115

Males over 14, 3,214; females over 14, 3,092; males and females under 14, 4,006

Source: Census of Lower Canada 1831.

mum expansion, the effects of a flourishing staple trade should have been visible: some elements of diversification might reasonably be expected to have taken place. As a minimal assumption, some agricultural production to feed the growing population was likely; it could also be anticipated that timber, as a supplementary staple growing out of the fishery's requirements for stages, flakes, stores, dwellings, and vessels of various sizes, would likewise develop to a degree. In the jargon of export-base theory, this would demonstrate the generation of some forward, backward, and final-demand linkages arising out of the fish staple itself.

Forward linkages may be defined as "a measure of the inducement to invest in industries using the output of the export industry as an input,"[9] that is, they reflect the possibility of processing the raw staple to an increasing degree. They are important because they add value to the staple at source, and thus provide additional income for the domestic economy. In comparison with other nineteenth-century colonial staples, the outputs of the fishery required relatively little processing beyond the curing and drying of the fish, so the flakes and assorted storage facilities on the beaches of Gaspé were all that was to be expected from the fish staple under the technological requirements of the age. Put another way, the forward-linkage effects of the staple were weak, although the dry cure used by Newfoundland merchants and by CRC was an improvement on the green cure used almost exclusively by France, Spain, and Portugal in an earlier age. That cure had entailed heavy salting down of the fish onboard vessels which caught the fish at sea and then processing it back in the mother country, thereby generating absolutely no forward-linkage effects on the Canadian side of the Atlantic. That old fishery had, in fact, been purely exploitative and had engendered no settlement. The forward linkages of the fish staple in Gaspé did generate settlement, but their effects were otherwise minimal.

Backward linkage – defined as "a measure of the inducement to invest in the home-production of inputs, including capital goods, for the expanding export sector"[10] – is much more complex to deal with. The production equipment of the fishery, excluding the vessels for the time being, was simple: stages, nets, and associated fishing gear for the most part. However, while stages were obviously built *in situ*, nets and gear were imported by the merchant as supplies into the fishery (see chapters 2 and 6).

Transportation is probably the most important backward linkage in a staple economy, since it lays the basis for a space economy as opposed to point development in an area. But landward transportation development in the fishery was virtually nonexistent. The catching and collection of the staple took place from the littoral, and it was processed on or close to the beaches – there was no reason to develop an extensive landward transportation network. Nor was there any reason to penetrate much of the hinterland or to clear it. The most clearance anticipated in the fishery in Gaspé was the nine miles back from the coastline estimated by Charles Robin as necessary for procuring timber for his shipbuilding needs.[11] Indeed, the dearth of roads in Gaspé prompted one observer, when asked about "communication deficiencies" in Gaspé County in 1833, to exclaim, "I don't understand this Question. So far as respects the County of Gaspé there are no roads of communication which can be called perfect by any possible construction of the term."[12] Pierre Fortin, the Canadian Fisheries Officer on the coast in the 1850s and 60s, understood only too well, however. Twenty-five years later he wrote of Gaspé that "le manque absolu de chemins a empêché jusqu'à présent les inhabitants de la côte d'aller s'établir dans l'interieur où les terres sont unies, d'un sol excellent, et couvertes de plus beaux bois."[13] George LeBoutillier, son of a Jersey merchant, summed up the situation when he wrote in 1860 of the proposed colonization roads that "settlement and trade will obtain ... an increase of grain and produce ... Produce is always at higher prices than those of the dearest markets in the Province ... In a word, our country imports all the produce of agriculture while it exports none. But we hope that the time is not too far distant when we shall raise all these products on our soil and when our fisheries shall be only a secondary branch of industry; a time when we shall practically realise the richness of our country, and for this improvement, we shall be, for the most part, indebted to the Colonisation Roads."[14]

The cod fishery required no roads, since its communication links were the sea lanes, and so its merchants built none. The fishing village, as figure 17 shows, represented a complete functional integration and spatial concentration of the industry on the littoral. This diagrammatic representation of Caraquet is typical of the layout of a merchant-fishery village. Caraquet (an important Robin station across the Baie des Chaleurs from Paspébiac) and other similar merchant-fishery villages were not like l'Anse au Loup, Labrador, or Point Lance on the Cape Shore of Newfoundland, where different households had their own waterfront access and their own household fishing economy.[15] Rather, Caraquet and its ilk were really industrial complexes serviced by a resident labour force as well as by a transient one (many of whom stayed in the merchant premises – "Batchelor's Hall"), and the physical layout of such villages em-

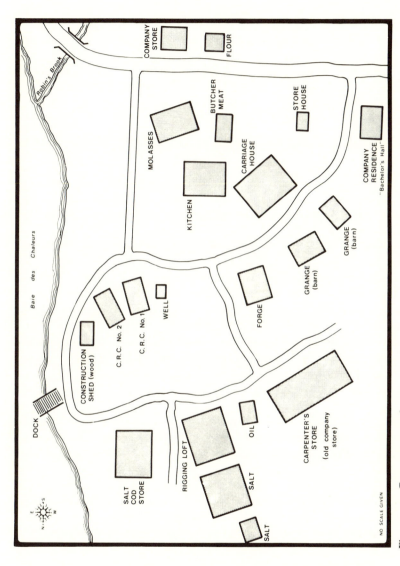

Figure 17. Caraquet, c. 1840

phasizes this. The village *was* the industry: the productive unit, the processing unit, the point of collection, the point of distribution, the point of importation, the point of exportation, and the point of control (locally) over all these functions. It was small and compact, and it looked to the sea for the exploitation of the resource which was its *raison d'être* and for which it was the point of access. In the days of the old migratory-ship fishery, access to the resource and to production and transportation facilities had all been subsumed within the vessel; today, the factory ship fulfills the same function within a new technological framework. A resident merchant fishery, however, meant the need to control settlers on a land base which was thought of as a sea-facing and sea-oriented production base.

One other aspect of transportation, however, was important to the Jerseymen: shipbuilding. Charles Robin wrote to Jersey in 1801 that "Mr. Day's gang is now preparing the timbers for the frame of the sharp vessel … As soon as that's done we mean to put up one like the others, having all the timbers ready prepared."[16] The firm cut and finished its own local wood: "Mr. Day has now [1795] nearly 200 pces. of timber cut which he is siding and squaring, the latter being performed on the Logs, keelpieces and such. The main road is all made and after tomorrow John Robin begins to superintend a gang which will be making the cross roads fr. each piece to the main road. Afterwards as soon as there is six to nine inches of snow, our drawing will take place. Some of our timber is four miles from us."[17] Shipbuilding should, in theory, have generated its own set of linkages, but, beyond the cutting of timber, very little of that developed in Gaspé.[18] The industry was actually more important to Jersey than to the fishery, and that explains why there were so few spin-off linkages from it to be found in Gaspé – they were developed at the metropole (see chapter 6). Indeed, the Gaspé-built vessels contributed significantly to the total Jersey home fleet, with forty-six New World-built vessels registered in Jersey in 1830, 29% of the total fleet for that year. Of this 29%, 21% was built at Paspébiac, 25% elsewhere in the Baie des Chaleurs and Gaspé, and 21% at Cape Breton establishments. Total New World tonnage registered in Jersey in 1800–1813 amounted to 1,400 tons (ten vessels) old measurement. Three thousand tons (twenty-five vessels) was registered between 1814 and 1823, 3,200 tons (twenty-five vessels) between 1824 and 1833, and 1,300 tons (twelve vessels) between 1833 and 1840.[19] In the 1830s, however, as preferential duties on Canadian timber were reduced and then removed, the Gaspé industry declined and production was transferred to Jersey. The existence of two sawmills in the Paspébiac area is the only real evidence of backward-linkage formation out of the fishery; it is to be associated with the various forms of construction required, of which shipbuilding was one prime example, stores, stages, and dwellings the other.

The lack of landward backward linkages out of the fishery hindered hinterland development, while shipbuilding tended to reinforce the ocean transportation structures that facilitated the links between each small New World locale and the mother country. While it cannot be assumed that roads would have automatically stimulated diversification of the economy or created regional links, it can be argued that without them even the preconditions for domestic development were not being achieved. Vance, in his mercantile model of endogenic and exogenic forces in newly settled areas, has shown the development of a transportation system from what he terms the point of attachment between two continents and the hinterland which the new area develops.[20] He describes this system as "entrepreneurial tentacles" or "transportation-trading links." In drawing attention to the inadequacy of central-place models to deal with early colonial-mercantile expansion in the New World, he emphasizes the importance of the exogenic nature of the colonial system and the development of trading foci in the hinterland that were to become entrepôts for staple collection. But he fails to take into account the case in which hinterland depots for staple collection will not develop. In this case, the development of a hinterland central-place system, servicing the staple area, will also not develop, since the fishing village subsumes all such functions. Moreover, in a colonially controlled staple trade, with an import-export monopoly residing in mercantile hands, the fishing village has no need of central-place services, nor of a hinterland, since it can bring in by sea all its needs and export all its produce without drawing upon the surrounding area. Put another way, the fishery can assemble both its inputs and its outputs with no additional transportation costs. There is, then, no "articulation point," since the "consumption system receiving trade" and the "production system returning trade" are subsumed within the same outport.[21]

What features of central place did develop did so only insofar as they related to staple merchant functions. What evolved in the merchant fishery was a means of relating continents on opposite sides of an ocean, and the focal point – the "point of attachment" – occurred on the littoral. In fact, the littoral consisted of a chain of points of attachment linked by chains of coastal transportation, and the wonderfully controllable character of such coastal stations was a major source of merchant monopolistic power. Arguably, in the case of a merchant fishery, the central-place system evolved as a littoral membrane or chain, as figure 18 demonstrates.[22] Paspébiac, for example, acted as the principal collection depot for the fish shown in the figure, and as a co-ordinating headquarters for the whole industry. The larger establishments acquired this function to a lesser degree under Paspébiac's central control, assembling inputs and outputs for the small stations along the coast. Figure 18 is merely a stylized description of CRC's business. Thus, Fortin would say in 1861 that "Paspébiac displayed on every side unmistakeable signs of commercial activity indicating its claim to be considered as the business centre of the Baie des Chaleurs."[23] But it was a restricted centre, looking overseas to Jersey and along the coastline of the Gaspé. Its function as a production centre was dominant, and its service functions minimal.

The third kind of linkage formation in a staple industry is that of final demand – "a measure of the inducement to invest in domestic industries producing consumer goods for factors in the export sector."[24] Obviously, the size of the domestic market is important here, and it in turn is dependent on the distribution of domestic income, and on how much of this income accrues to the core (here, the mother country). Basically, as Baldwin, North, Watkins, and Gilmour all point out, if per-capita income is high and equally distributed, then final-demand linkages will be strong, since the presence of consumers will stimulate local production of goods and services.[25] If, however, per-capita income is low, subsistence (home production) will usually follow; if per-capita income is unequally distributed, then luxury imports will be in demand at the higher-income level, while subsistence will predominate in the low-income group.

The simplest form of final-demand creation in Gaspé would have been the establishment of agriculture, at least to the point at which the region would be self-sufficient in foodstuffs that could be grown locally. In fact, agricultural development was extremely limited. Improved acreage amounted to only 17.4% of all occupied land in Gaspé County and 12.3% in Bonaventure County, and, such as it was, it was all close to shore – no development of an agricultural hinterland occurred. In 1831, the returning officer for the *Census* reported of Gaspé, "This County is entirely dependant upon the Fisheries. A few traders, labourers, farmers and planters reside at Little River, Cape Cove, L'Ance au Beaufils, Percé, Long Beach, Mall Bay and Point St. Peter. The N.W. and s.w. arms of Gaspé are all farmers ..."[26] In 1833, two years after the *Census* was taken, the sub-collector of customs at Gaspé reported that the "front is occupied wherever it gives a convenient access to the water for the purpose of carrying on fishery although in most cases very small backwards

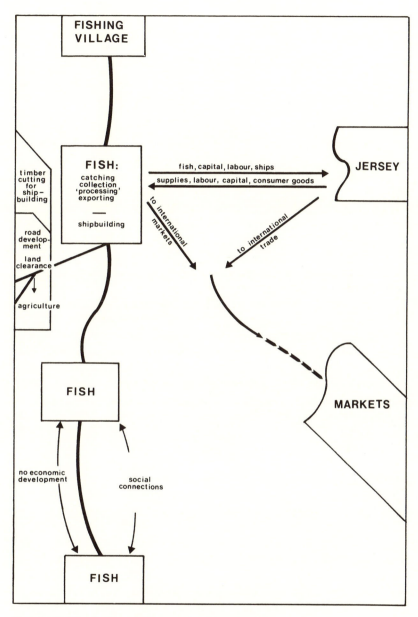

Figure 18. The Fishing Village as a Jersey Outpost, c. 1830

Table 10
Subsistence Earnings from Agriculture in Gaspé, 1831

	Percentage of Households	Percentage of Population	Percentage of Farm Servants
Gaspé	0.8	0.14	
Bonaventure	48.8	8.6	6.0

Source: Calculated from the 1831 Census.

be cleared ... I am not aware of the 2nd concession being occupied anywhere in this County." Only three persons in Gaspé County were involved in farming as a sole pursuit – one in Gaspé Basin and two in contiguous s.w. Arm – and agriculture was clearly practised as a support for the local population in terms of fulfilling need, not as a commercial sector of the economy. Indeed, the sub-collector's *Report* was explicit about this, noting, "The inhabitants generally hold lots of land and employ their leisure time in Agricultural pursuits, cultivating potatoes, wheat, barley, oats and hay without which I am of the opinion they could not support themselves through a long winter." Farming was really an adjunct to the fishery: "Fishing and farming may be carried on by the same family with more success than if either pursuit were followed distinct, as there are many days during the fishing season not adapted to that pursuit when the Fishermen can be profitably employed in Farming."[27]

Bonaventure County seems to have been rather better off than Gaspé County with respect to agricultural practices. Table 10 shows that while a large proportion of the population of Bonaventure County gained some earnings from agriculture production, such earnings were virtually nonexistent in Gaspé County. However, a breakdown of agricultural production per household (table 11) clearly indicates the subsistence nature of farm production even in Bonaventure County, although the situation was more extreme in the fishing area around Paspébiac (Bonaventure) than it was at the head of the bay (Restigouche), with a heavier dependence per household on potatoes, oats, and cattle. Potatoes were the basic subsistence crop grown in the area; much (though not all) of the flour consumed locally was imported by CRC and sold in the company stores. Dairy cattle likewise were kept for local consumption, and McConnell noted specifically that "milk is not an article of merchandise, the sheep and the calves get the surplus skim milk and the pigs are treated to that agreeable beverage Buttermilk."[28] The *Census* for Gaspé County provides enough detail to permit analysis

Table 11
Agricultural Production per Annum (in Minots)

	Bonaventure County	Restigouche Area	Bonaventure Area	Gaspé County
Wheat	5,470	3,850	1,620	4,872
Peas	432	342	90	488
Oats	3,600	1,000	2,600	1,920
Barley	3,400	2,500	900	1,583
Rye	16	16	0	302
Potatoes	426,940	159,140	264,060	102,525
Neat cattle	3,195	1,525	1,620	2,216
Horses	360	230	130	317
Sheep	5,318	3,106	2,210	3,662
Hogs	3,971	2,421	1,400	2,438
Number of Minots Produced per Household				
Wheat	5.8	8.5	3.95	5.6
Peas	0.5	0.75	0.22	0.56
Oats	3.8	2.2	6.3	2.2
Barley	3.6	5.5	2.2	1.8
Rye	–	–	–	0.4
Potatoes	454.7	351.3	644	118.5
Cattle	3.4	3.4	4.0	2.6
Horses	0.38	0.5	0.3	0.37
Sheep	5.7	6.9	5.4	4.2
Hogs	4.2	5.3	3.4	2.8
Number of households	939	453	410	865

Source: Calculated from the 1831 *Census*.

of the Robin seigneurie of Pabos and Grand River, where the influence of a fishery on agriculture is clearly visible. In this seigneurie, the figures for average household production of wheat (0.6 barrels), of potatoes (125 barrels), and of cattle (1.4 animals) were all significantly lower than even the lowest figures in table 11.

Clearly, the fishery encouraged a heavily subsistent village economy in those areas of Gaspé in which it was dominant, with extreme reliance placed on the potato and all else reduced to a minimum – this despite the fact that there was some agricultural potential in the area, as table 11 indicates and George LeBoutillier observed. Samson's observation that "farm work was done and community services provided when fishing was poor"[29] makes sense in this context, particularly when the temporal incompatibility of seed-time and the fishing seasonal round (chapter 4) is recalled. The point is that agriculture, however restricted the local potential for such might have been – and any real assessment of that potential would require a detailed study of the Gaspé climate and soil quality – was not seen by the fish merchant as something to be encouraged.[30]

Given the slight development of the agricultural base in the area, not much growth could be expected in related sectors of the economy. Indeed, in Gaspé County there was only one gristmill for the whole county, and there was no other processing or manufacturing. In Bonaventure County, again slightly better off, there were five gristmills (1:1,061 head of population) and three sawmills (1:1,770 head of population). One of each was situated at the head of the bay, where there was some farming activity, and the other four gristmills (and two sawmills) were in the fishing area around Paspébiac. Given the need for inexpensive flour in the fishing establishments, it is not unlikely that the presence of CRC would stimulate a demand for gristmills, located here rather than in the producing area further up the bay. The Gaspé mill was situated in Gaspé Basin, in the only area where farming was carried on independently of the fishery, and it was small and owned by the only "extensive farmer," James Stewart of Gaspé Basin.

An analysis of occupational structure in Gaspé underlines the dominance of the staple sector. Table 12 shows selected occupations in the Gaspé County economy for 1833; the categories are not mutually exclusive.[31] In Grand Grève, for example, the firm of P. and F. Janvrin was at the same time the fish merchant, the ship owner, and the owner of the fishing establishment. Gaspé Basin was clearly the administrative centre of the county, as it included the residence of the justice of the peace, the collector of customs, members of the militia, and the minister (Church of England). The number of fishing establishments is underrepresented (thirty-two enumerated) – indeed, fishermen themselves are not even mentioned – and many more "small" establishments were noted in the *Report* but not counted. Most of these smaller establishments were owned by Jerseymen, as outposts of their main establishments – CRC's operation at Percé, for example, of which firm McConnell remarked that they were "merchant shipowners and carrying on fishing in the most extensive scale." The fishing stations far outweighed any farming operation in the area, of which all but three were carried on in conjunction with the fishery or fish-related occupations. For example, boatbuilding was carried on thus at L'Ance au Cousins, as was shipbuilding at Malbaie – by the Mabé family (three males) the shipbuilding firm that built the Perrée vessel whose sale was detailed

Table 12
Occupations in Gaspé County, 1833

	"Officials"[1]	Church Men	Merchant (Fish)	Farmer	Grist Mill	Whaler	Shipbuilder	Blacksmith	Merchant (Lumber)	Ship Owner	Fishing Establishment	Trader
Gaspé Basin	4	1										
South West Arm			1	4		1	1					
L'Anse aux Cousins				2			2					
Sandy Beach			1								1	
Peninsula						3						
Douglastown									1			
Grand Grève			1J[2]							1J	1J	
St George's Cove			1J				1J			1J	3JJ	
Pt St Peter	2		4JJ						1	2JJ	3JJ	1
Mal Bay			2	3	1		1	1			6J	
Percé			3JJJ				1		1	1J	6JJJ + many small ones	1
Cape Cove			3JJ								3JJ	
Gd River			2JJ								2JJ	
Anse Griffon											2	
Fox River											4	
Grand Étang											1J	
Totals	6	1	18 all 11J	10	1	4	5 all 1J	1	2	5J	32 all, 16J + small ones	2

Source: Baddely's Report

Notes: [1] Justice of Peace, Captain Royal Marines, Collector of Customs

[2] J = Jersey-owned

Table 13
Gaspé Shipping for the County, 1831

Rig	Name	Tonnage	Owner	Established in area	Place of Origin
Ship	*Messenger*	247	Janvrin	Yes	Jersey
Ship	*Janvrin*	228	Janvrin	Yes	Jersey
Brig	*Doris*	168	Janvrin	Yes	Jersey
Ship	*O. Blanchard*	?	CRC[1]	Yes	Jersey
Brig	*Broad Axe*	?	CRC[1]	Yes	Jersey
Schr-Brig	*Dit-on*	?	CRC[1]	Yes	Jersey
Brig	*D'Amour*	141	J. Perrée	Yes	Jersey
Brig	*Adventure*	113	J. Perrée	Yes	Jersey
Brig	*Canada*	144	W. Alexandre	Yes	Jersey
Schr.	*Minnow*	63	J. Vibert	Yes	Jersey
Schr.	*Habnab*	138	J. Vibert	Yes	Jersey
Schr-Brig	*Prince Regent*	79	J. Rossier	Yes	Jersey
Schr-Brig	*Concord*	86	J. Rossier	Yes	Jersey
Schr.	*Spartan*	57	E. Le Rossignol	Yes	Jersey
Schr-Brig	*Superb*	86	J. LeBoutillier	–[2]	Jersey
Brig	*St E.*	133	Monamy and Ahier	Yes	Jersey
Schr-Brig	*Susan*	87	Le Gresley & Co.	–	Jersey
Brig	*Commerce*	112	Le Gresley & Vibert	Yes	Jersey
Schr-Brig	*Young Peggy*	60	Le Cornu	–	Jersey
Brig	*Adonis*	142	F. Bertram	–	Jersey
Schr-Brig	*Judith and Esther*	83	J. Vibert & Co.	Yes	Jersey
Brig	*Seneca*	103	P. Perchard	Yes	Jersey
Brig	*Egton*	63	P. Duval & Co.	Yes	Jersey
Brig	*Friends*	115	J. Vibert & Co.	Yes	Jersey

Source: Baddely's *Report*; CRC Letterbooks; "Nouvelles de Mer" of the *Chronique de Jersey*, Jersey Shipping Registers 1803–33.
Notes: [1] CRC had other ships (see Letterbooks), but these landed goods in Bonaventure County.
[2] – Not given.

in chapter 2. The impression given by the table is one of over-whelming dependence on the fisheries, through its merchants and establishments; the predominance of the Jersey merchants is also obvious. What "development" did occur was either in the hands of merchants or (as with the Mabé family) in the hands of people working under contract for the Jerseymen. Indeed, there was total Jersey control of ship-owning (see table 13) in Gaspé County, and this meant that the Jersey merchants effectively monopolized the import/export trade of the area.[32]

Nor did there appear to be much chance, in 1831, of the local population creating any effective demand which would have stimulated further growth, despite a rapidly expanding population base, because of persistently low per-capita incomes, which were almost ubiquitous in the Gaspé cod economy; even the merchants from Jersey were parsimonious in their expenditure of capital in the area.[33] The number of persons in higher-income groups was insignificant, and they therefore did not provide a focus for incipient domestic production. Even some subsistence commodities, such as coarse wearing apparel, had to be imported into Gaspé.

This is not hard to explain. The preferred mode of conducting business with residents of the area was barter. The merchants imported and sold those articles needed for the fishery along with such items as the area could not produce for itself at the subsistence level – "wearing apparel, provisions, Rum, Sugar, Molasses, tobacco, tea, coffee, rice, Indian meal, etc.," as McConnell listed it. Wages were paid "truck,"[34] and those who fished were either on "half their catch" or on "store payment," which he described as being "completed at about 25% premium which (the risk, expenses of handling etc. in receiving fish payment taken into account) is by no means an extravagant or too liberal a difference." He added that the "operative fisherman" had to possess "hooks, lines, boat, provisions," and therefore, unless he was "very attentive and sparing," would have "very little to his credit at the end of the season"[35] – which meant that his capacity to purchase anything beyond necessities would be severely limited.

McConnell's qualitative assessment suggests that the poverty of the area was a result of the local population's inability to accumulate capital because of the way in which the merchant credit system operated. This can be seen from the ledgers of the firm. From 1826 to 1850 (table 14), there was an average of about five hundred active accounts annually in the books of the two stations combined; after 1850 that figure increased rapidly, so that by 1862 there were nearly one thousand clients between the two stations.[36] The rise in the number of accounts parallels the rise in production by the firm (table 8). At the Percé station, zero-balance accounts (no debit or credit on the books) were most numerous throughout the period, positive accounts (a credit on the books) least numerous, and negative accounts (a debit on the books) declined steadily after 1838 from 45% of all active accounts to 12% by 1856. Thereafter, negative accounts rose, while zero and positive accounts declined, suggesting an increase in indebtedness to the firm. In Paspébiac the picture was more complex, since the headquarters handled more and more of the British North American dealings of CRC as the century wore on. Zero accounts were never so dominant at Paspébiac; positive accounts never outweighed them; and negative accounts were dom-

Table 14
Balances: Paspébiac and Percé, 1826-64

	Paspébiac				Percé			
	per cent				per cent			
Year	+	−	0	No.	+	−	0	No.
1826	18	56	26	285	9	36	55	217
1828	14	67	19	284	11	46	42	246
1830	7	67	26	279	8	56	36	224
1832	11	70	19	208	6	49	45	290
1834	11	51	38	368	7	53	40	188
1836	23	33	45	316	7	36	57	202
1838	14	44	42	327	7	45	49	155
1840	17	45	38	265	18	31	52	194
1842	29	31	40	255	22	28	50	199
1844	19	35	46	284	19	23	58	233
1846	25	25	50	274	26	25	57	240
1848	24	47	29	264	23	21	56	252
1850	24	42	33	238	24	21	55	246
1852	23	43	34	236	43	15	42	233
1854	34	44	22	254	32	14	53	247
1856	36	49	15	276	25	12	63	282
1858	33	36	31	344	22	18	61	316
1860	28	38	34	346	23	20	56	393
1862	23	47	30	386	16	35	49	517
1864	32	40	29	461	nd	nd	nd	nd

Source: Robin, Jones and Whitman Papers, NA, MG 28 III, vols. 172–9, 181–3, 185–6.

inant for almost the whole period, ranging from a high of 73% in 1829 to a low of 25% in 1846.

In Gaspé, CRC – while always "debit-led" (that is, the firm normally was owed more than it owed) – was less so during the 1830s and 1840s, but increasingly so after the Reciprocity Treaty of 1854. Over the whole period, the firm never owed its clients more than £4,202 at Paspébiac or £1,712 at Percé in any given year and, before 1846, never more than £1,000 at either station. By the same token, the firm was owed as much as £7,615 at Paspébiac and £3,623 at Percé, although, as table 15 shows, the overall picture was one of declining client indebtedness from 1834 to 1850,[37] and then a rapid rise in client debt after 1854. Until the late 1840s and then again after the mid-1850s, the value of client debts to the firm outweighed the value of its debts to them, as table 16 shows, but in the intervening period CRC was relatively more indebted than its clients (asterisked figures

Table 15
Increase/Decrease in Amount Owed by/to CRC,
1828–62 (Pounds Sterling, Paspébiac and
Percé Combined)

Year	Firm Owes	+ / −	Firm is Owed	+ / −
1828	30	−	731	+
1830	335	−	451	−
1832	112	−	392	−
1834	714	+	751	+
1836	349	−	1,809	−
1838	6	+	559	+
1840	102	+	942	−
1842	918	+	954	−
1844	7	−	8	−
1846	386	+	443	−
1848	194	+	314	+
1850	111	−	297	−
1852	671	+	502	−
1854	187	+	237	+
1856	86	+	955	+
1858	249	+	915	+
1860	376	+	1,782	+
1862	458	−	4,898	+

Source: Robin, Jones and Whitman Papers, NA, MG III,
vols. 172–9, 181–3, 185–6.
Note: + / − is increase (+) or decrease (−) from previous
year shown; figures rounded to the nearest pound.

on table 16). These data suggest that CRC, secure in its established
monopsony on the coast, had been in the process of developing
beyond its early reliance on client debt until the Reciprocity Treaty
upset the *status quo* by opening the British North American inshore
fisheries to the United States, thus leaving CRC vulnerable to American traders willing to pay cash for fish.[38] Certainly there was a
reduction of client debt between 1846 and 1856, but whether it was
a result of the firm getting stingier with its credit or clients being
able to pay more to the firm is not revealed by the aggregate figures.
By 1856, with Reciprocity in place, the old scenario of client indebtedness had reasserted itself.[39]

Restriction of cash can be clearly demonstrated from the account
books. CRC's ledgers show few cash transactions in the early years:
from 1833 to 1847, the average cash transactions per annum
amounted to only £2,879 for all stations; in the later years (1858–
63), that amount rose considerably, to an average of £4,940

Table 16
Value of CRC's Debts as a Percentage of Client Debts,
1826–64

Year	Paspébiac	Percé
1826	38	21
1828	22	10
1830	5	7
1832	4	7
1834	23	8
1836	25	5
1838	19	5
1840	28	10
1842	92	41
1844	87	41
1846	131*	59
1848	95	65
1850	114*	60
1852	99	201*
1854	117*	167*
1856	82	140*
1858	80	87
1860	62	66
1862	28	34
1864	55	n.d.

Source: Robin, Jones and Whitman Papers, NA, MG 28 III,
vols. 172–9, 181–3, 185–6.
Note: One hundred per cent is "break-even," when the firm's
debts equal the clients' debts. Below 100 per cent, the firm's
debts to its clients are less than client debts to the firm;
above 100 per cent, the debts of the firm are greater than
those of its clients.

per annum; and in the intervening decade (1848–57), the cash account was severely restricted – averaging only £1,964 per annum at
a time when cash was being used on the coast by American and
other traders. These figures suggest deliberate stringency on the
part of CRC – in line with its general retrenchment during those years
– and a deliberate preference for credit. Indeed, throughout the
ledgers, cash transactions for any substantial amount were almost
never with fishermen, but with other firms or with people whom
the firm did not deal with on a regular basis. As such, they may
reflect something of the wider contacts of the firm in the area, but
they tell us little about the ongoing relationship between CRC and
its usual clients. Cash transactions with fishermen were customarily
in the form of cash "loans" for less than £5, before 1855, and almost

never for more than £50, even in the later years. Often these were not really cash transactions at all but merely transactions on the books, that is, a debt to a certain value, transferred and paid through an account.[40]

Cash price and truck price for goods were, of course, related, since the truck price was the cash price plus a premium charged by the firm, often at very high rates: 25% to 40% was usual, and even higher rates (up to 100%) were not uncommon.[41] It is the level of these premiums, rather than simple indebtedness, that defines CRC's credit system as clearly reducing the fisherman's real wage below his nominal wage and thus deliberately restricting incomes on the coast.

To prove that CRC was debit-led and operated a truck system is one thing; to get some idea of how that system affected people's lives is a very different matter. The kind of internal evidence that can be extracted from individual accounts, and especially for several individual clients over a period of time, puts a human face on the statistical analysis for which these kinds of data are more normally used.[42] Table 17 shows one such account by way of illustration. This is the account, for one year only, of a fisherman, one of thousands of similar format. On the left-hand side is the amount owed to the firm. It is made up of the previous year's unpaid balance, money paid by the firm into a sharesman's account (that is, that person's share with the account-holder of the fishing that he did over a season), a debt owed to the brother of the account-holder and credited by the firm to the brother's account, and "sundries," which is the account-holder's debt to the company store. The ledgers do not detail "sundries," but McConnell described store inventories as basic necessities such as "salt, netts, lines, hooks and other fishing tackle, British manufactured dry goods … also cider, cider vinegar … canvas, cordage, ironwork, hardware and fittings for new vessels. Coarse clothing is one item much in demand. Flour, salted provisions, rum, sugar, molasses, tobacco, tea and other groceries are imported."[43] On the right-hand side is what has been paid by the account-holder toward his debt. It is made up of sugar (which he must have been returning to the store), fish, money from his brother, and "days work": wage labour of an unspecified nature. The account shows several grades of fish: "green" (uncured), the cheapest; "inferior" (poor cure); and "fish" (good cure, marketable), usually earning the highest price. "By balance due" is the amount still outstanding at year's end, which would be transferred to the following year's account and placed on the left-hand side as a debt to the company. The total amount was for £76–3–3.

Table 17
Account of Rhéné Dugay, 1826

Date	Per	£ – s – d	Date	Contra	£ – s – d
	Bal. last year	$48 - 15 - 3\frac{3}{4}$	May 4	By 72 sugar @ 6d	$1 - 16 - 0$
June 10	Cred. fo. LeBlanc his share fish	$1 - 14 - 10\frac{1}{2}$	June 10/28	By green fish $97\frac{3}{4}$ qtls @ 5/-	$24 - 8 - 9$
Sep. 12	To Bon to bro. Frederick	$1 - 0 - 0$	Oct. 17	By $7\frac{1}{2}$ days work @ 3/6 per Frederick	$1 - 6 - 3$
	To Sundries per a/c	$24 - 13 - 0\frac{3}{4}$		By 2 " 1 " 14 Fish @ 12/6	$1 - 9 - 8\frac{1}{4}$
				By 0 " 0 " 21 Inf. do. @ 10/-	$1 - 10\frac{1}{2}$
				By $3\frac{1}{2}$ days work @ 3/6	$6 - 1\frac{1}{2}$
				By Bal. due. Carrd. to fo. 63	$46 - 14 - 6\frac{3}{4}$
		$76 - 3 - 3$			$76 - 3 - 3$

Source: Robin, Jones and Whitman Papers, MG 28 III, vol. 173.

Analysis of the account, then, reveals that this person was capable of being a good fisherman: he had marketable fish to his account. He was a sharesman of some description: he crewed with LeBlanc for a share of the catch they both made. He was in debt: he could not make enough on fishing alone to pay his account at the store. He did reduce his debt, but only by virtue of "days work" and returning sugar to the company. He was not a "bad debt": he was not "cut off" by the firm, which is to say that his company account was continued, but he was in debt. Thus, a careful reading of the internal evidence contained in individual accounts can provide a great deal of information, especially if such accounts are followed over an extended period of time. The occupation of clients can usually be elicited, as can whether or not a client was employed, retired, ill, or dying (the nature of payments to the firm often indicates this); of course, long-term financial status can be ascertained by tracking individual accounts over their lifetime on the books. However, when the number of clients on CRC's books is considered, it becomes obvious that only computerization of the thousands of individual accounts involved will ultimately allow the full analytical potential of such records to be realized. In the meantime, the alternative of using illustrative case studies of the kinds of situations in which CRC's clients found themselves has been adopted here, with only fishermen's accounts being used since they can safely be taken to be generally representative of the bulk of CRC's clientele.

Table 18 shows the summary histories of thirteen accounts taken from the early phase of CRC's operations (1826–52); table 19 shows an extended family's account history over the middle and later years (1842–77).[44] These examples are drawn from the Paspébiac ledger, but the same kind of information is to be found throughout the books of all the stations. The purpose here is limited: to ask what the ledgers reveal about how individuals on the coast were dealt with under the merchant-credit system over time. The first impression gained from reviewing the accounts is one of enormous variety: the system was not monolithic, with everyone faring equally badly, being equally trapped or equally exploited. Some fishermen never went into debt (Sire and Lantin), one only rarely (Doiron), while others were eventually cut off and were not permitted to make purchases at the company store (Hébert, Castillon, and Dugay). The Arsenault family showed the same kind of variation in the later years, though none of them managed to escape debt eventually and four were cut off. Clearly, the system was more complex than the mythology about it has acknowledged to date; the reality was far from monochromatic.

Table 18
Thirteen Case Studies: Individual Accounts

Year	Arsenault	Doiron	E. Huard	Hébert	Lantin	J. Huard	Chapados	Arosbile	Holme	Sire	Hautseinnet	Castillon	Dugay	Number + balances
1826	+	−	0	−	+	0	+	−	−	+	0	−	−	7
1827	0	0	0	0	0	−	−	−	−	+	−	−	−	3
1828	+	0	+	−	0	−	−	−	−	+	−	−	−	6
1829	0	0	0	0	0	−	−	−	−	+	−	−	−	3
1830	0	0	?	0	0	−	−	−	−	0	−	−	X	2
1831	−	0	0	0	0	0	−	−	−	+	0	−	−	3
1832	0	−	−	−	0	0	−	−	−	+	0	−	−	3
1833	−	0	0	−	0	0	−	−	−	+	0	−	−	3
1834	−	0	0	+	0	0	−	−	−	0	+	−	X	1
1835	0	0	0	+	+	−	−	+	0	+	+	−	X	7
1836	0	0	0	0	0	−	−	0	0	0	0	−	−	3
1837	+	0	0	+	0	−	−	0	−	+	+	−	−	2
1838	−	0	0	+	0	−	−	−	−	0	0	−	−	0
1839	+	0	0	+		−	−	−	−			−	−	3
1840	−		−	+		−	−	−	0			−	−	1
1841	−		−	0		−	−	0		+	+	−	X	1
1842	0	0	0	−		0	−	0	0	+	+	−	X	2
1843	0			0			−	−		−	−	−	X	1
1844	0			0			−			−	+	−		1
1845	+			0			+			−	+	−		2
1846	0			0			0		0	−	+	−		3
1847	−			−			−			−	0	−		2
1848	−			X			−					−		0
1849	−	ç		X			−					−		0
1850	−			X			0		0			−		0
1851				X								−		0
1852				X								X		0 →

Left-margin brackets: **bad Fishery** (1828–1830); **bad Fishery** (1837–1840); **poor Fishery** (1850–1851).

Annotations: Lantin — ——— Moves to Caraquet ———→ ; Sire — ——— Moves to Grand River ———→ ; Dugay — ——— Dies ———→

Source: Robin, Jones and Whitman Papers, MG 28 III 18, vols. 173–92, passim.

Key: + = positive balance   0 = account balances   X = cut off, bad debt to 1856   ▼ = account closed
− = negative balance

Some clarification of how the system actually worked can be obtained from looking at the actual amounts of balances rather than at simply their status (positive, negative, or zero). The three fishermen who were cut off in the earlier years (table 18) had balances that ranged from -£11 to + £8 (Hébert), from -£6 to -£62 (Dugay) and from -£6 to -£40 (Castillon). In all of these cases, CRC kept them on the books for as long as they were making regular payments that reduced the outstanding amount even to a small degree. Castillon, for example, owed £40 on a year's total bill of £87 in 1842, and £38 on £81 in 1845, but in 1850 he owed £40 on £61, and when he still owed that in 1851, now on an account of £80, he was cut off and thus not allowed to incur more debt. Hébert usually paid the firm in fish or in fish and wood. In 1840 he supplemented that with farm produce. By 1848 he was unable to make any payment on the account and was cut off, his balance remaining on the books as a bad debt. Dugay (whose account for 1826 was shown in detail in table 17) was a similar case. He was cut off in 1830, and again from 1833 to 1836 and 1840 to 1842, because – despite efforts to reduce his outstanding balance through payment in fish, fish oil, wood, hay, return of purchases, and manual labour – he made little or no progress over a very long period. Even assistance from his son (a credit of £26–12–5 $\frac{1}{2}$ is shown as transferred from his son's account in 1833) was not enough to prevent his being cut off. Finally, in 1843, with only wood on the credit side of his account and a debt of £29–1–9, Dugay died, his balance remaining as a bad debt.

CRC was not a compassionate company; there is no evidence of charity in the account books. Not even the children of Philip Robin and Martha Arbou were assisted beyond an annual credit on the books for a small allowance.[45] The one possible instance of assistance given was with the account of James Day, master shipbuilder for Charles Robin, after Day's death in 1833. The next year, James Robin sent an order to the firm's London agents clearing the account of all but £3–18–11, which was a charge for the transactions involved and which was later paid by the church. It is also possible, of course, that Day's estate paid CRC, which then cleared the account.[46]

Table 19 looks at the later phase of CRC's operations through the accounts of one Acadian kin group. What is really significant in this set of accounts is the strong evidence of a weakening of indebtedness in the 1850s and then escalating debt in the 1860s. The accounts of the Arsenault family, considered alongside the increase in production (table 8) and in debt (tables 14 and 15), show the same general pattern pertaining for CRC's clients, if this family is typical, and there is no reason to assume that they are not. When wages and the prices

Table 19
La Famille Arsenault, 1842–77

Year	Frederick	Peter E.	Joseph	J.-B.	Zachary	Sebastion	Felix	Number + balances	
1842	+		+			0		2	Fred's a/c starts £18
1843	+		+			+		3	
1844	+		0			0		2	
1845	+		+			+		2	
1846	+		+			0		3	
1847	+		+			0		2/3	
1848	+		0			+	+	3	Felix @ £10
1849	+		0			0	+	2	
1850	+		0			0	+	2	
1851	+		+			+	+	4/4	
1852	+		+	+		+	+	5	
1853	+		+	+		+	+	5	
1854	+		+	0		+	+	4/5	
1855	+	–	+	–	+	+	+	5/7	
1856	+	–	+	–	–	+	+	4	
1857	+	–	+	–	–	+	+	4	
1858	+		+	–	+	+	+	5	← best year for family
1859	+	–	+	–	?	+	+	4	Felix @ +£27
1860	+	n.d	–	+	–	+	–	3	Felix @ −£49
1861	+	cut off	0	+	–	+	+	4	
1862	+	–	+	cut off	–	+	–	3	Peter @ −£80; Felix @ −£401
1863	+	cut off	cut off	off	?	+	–	2	peak of Fred's a/c £152
1864	+	x	–	–	cut	+	–	2	

Year							Notes
1865	+	x	x	off	+	−	2	
1866	+	x	x	x	+	−	2	
1867	+	x	x	x	+	−	2	
1868	+	x	x	x	+	−	2	
1869	+	x	x	x	+	−	2	
1870	+	x	x	x	+	−	2	
1871	+	x	x	x	− dies; a/c to son →	−	1	Felix @ −£300
1872	+	x	x	x		−	1	
1873	+	x	x	x		−	1	
1874	+	x	x	x		−	1	
1875	+	x	x	x		−	1	
1876	−	−	−	−	−	−	0	Fred @ −£81; Peter @ −£55
1877	−	−	−	−	−	−	0	Fred @ −£6; Felix @ −£253

declining amount of positive balances

"the squeeze"

Source: Robin, Jones and Whitman Papers, MG 28 III 18, vols. 173–92, *passim.*

Key: + = positive   0 = account balances

      − = negative   x = cut off

      → = account closed

set for fish on the coast (set, that is, by the firm) are considered, it is easy to see how this could happen. Wages varied very little throughout the whole period. Work in the spring fishery in 1826 was paid at 2/6 per day, and at 3/6 in the summer fishery. In 1828 it was 2/6 and 4/- per day. By 1840 it was still 2/6, 3/6 and 3/9 per day, and by 1847, 4/- per day. Fish prices on the coast also remained stable, as table 20 shows, rising slightly in the bad years after 1836, only to return to previous levels until 1850. Indeed, right up to the late 1860s, these prices were very stable indeed and were clearly not set according to yearly changes in the market price for fish.

It is in the fine details of accounts such as these that the workings of the credit system at the individual level become clear. Full understanding of the system, however, must rest on a grasp of the reciprocity of trade-offs between merchant and client, along with an appreciation that the large merchant firm and the individual fisherman were not equal partners. In other words, the firm was always making judgments about who to keep and who not to keep. Its clients, of course, were making reciprocal judgments about how much to purchase without getting into the kind of debt that would result in their being cut off with nowhere else to go, since a bad debt with CRC was a bad risk for any other merchant on the coast.[47] For both merchant and client, good and bad years in the fishery had to be added into their calculations: good years were those in which local consumption could rise and the firm could expand its profits on supplies, while bad years meant that the firm had to tighten its credit and clients reduce their level of consumption accordingly. As the Jersey headquarters instructed Paspébiac in 1876, "Fear bad business ... you must be on your guard not to advance goods; if no fishery we must keep our goods, and only give out merely fishing articles; put the agents at Outposts on their guard, to be careful how they make advances."[48] The shift from a run of good years to a run of bad years was often a crisis point for individual accounts. Thus Arsenault, who had balanced his account in 1846 at £58–9–7, was in debt over the next three years to the tune of £1 to £4, while Hébert sank into irredeemable debt at the same time because he could make no inroads on the sum he owed. Accounts that the company considered viable were those that regularly showed a positive or zero balance, or payment on a negative balance (Arsenault, Doiron, Lantin, Sire); bad accounts were those whose owners could not provide enough fish, or alternatively enough produce or service, to keep a permissible small negative balance running. Such accounts often showed a scramble to keep from being cut off, replete with small quantities of produce ("sugar") or even private goods ("a bridle")

Table 20
Average Fish Prices, Market and "Coast," 1822–77
(Pounds Sterling)

Year	Market	Coast	Year	Market	Coast
1822	12/3	9/3	1850	11/9	10/3
1823	11/9	"	1851	12/–	"
1824	11/4	"	1852	16/6	"
1825	9/3	"	1853	13/9	"
1826	9/11	"	1854	15/3	"
1827	13/4	"	1855	16/6	11/9
1828	11/6	"	1856	15/3	"
1829	9/3	"	1857	14/9	"
1830	8/–	"	1858	14/9	"
1831	11/9	"	1859	15/10	"
1832	10/11	"	1860	15/6	"
1833	12/6	"	1861	13/9	11/3
1834	9/9	9/6	1862	15/–	"
1835	12/3	"	1863	18/11	12/–
1836	13/6	"	1864	19/–	"
1837	14/9	"	1865	17/3	"
1838	15/6	10/6	1866	22/6	"
1839	15/–	"	1867	15/6	"
1840	9/3	10/–	1868	14/9	"
1841	12/6	"	1869	16/10	"
1842	9/9	"	1870	16/10	16/–
1843	10/10	"	1871	16/10	15/–
1844	11/6	"	1872	16/10	"
1845	10/9	9/6	1873	17/–	"
1846	12/–	"	1874	17/6	"
1847	14/6	"	1875	19/–	"
1848	13/6	"	1876	22/9	"
1849	11/6	"	1877	18/9	"

Source: A. Bezanson, Wholesale Prices in Philadelphia,
1784–1861 (Philadelphia, 1936), 45; Ryan, Fish Out of Water,
264; Robin, Jones and Whitman Papers, NA, MG 28 III,
vols. 172–4, 177–92.

sold back to the store in an effort to reduce the amount outstanding by just enough to satisfy the firm. The crucial importance of subsistence agriculture, which served as a safety net in such circumstances, is highlighted by such entries as "potatoes" or "cabbage" being sold to the store to reduce outstanding debt, and manual labour (including housecleaning and laundry services) was also often entered as partial payment of an account.[49] It is also clear that anyone who failed to make payments on a debt for two consecutive years was liable to be cut off.

The question, of course, must be why local fishermen stayed with CRC under these conditions, since it is clear that they did. Indeed, the number of clients on the books rose steadily from 1854 onward, even during the crisis years of Reciprocity. One can only surmise that they stayed because they were able to increase their purchases on account with CRC at this time since the firm was concerned that it might find itself short of producers, and hence of fish. Of course, that rebounded on the clients later when their increased purchases became a source of indebtedness from which they could not escape once the threat of external competition had been overcome by the firm and it could recoup its losses to some degree. It is also worth remembering that CRC could always offer its clients something that the itinerant trader could not – a reliable source of supplies, regardless of good or bad fishing seasons. Such stability was vitally important to fishermen.

What this means, then, is that there was little likelihood of local capital accumulation on the coast while the truck system remained in place since, in effect, a cash surplus to create demand did not exist. Income was uniformly low, and even the Jerseymen lived frugally. As a result, no focal point was likely to develop as a strong market centre, not even Gaspé Basin, which lacked a sufficiently high concentration of high-income residents to create a significant demand structure (see table 12). It thus seems that the lack of diversification around the staple base in Gaspé was, to a significant extent, the result of a mercantile "industrial" venture which was solely concerned with its export trade and brooked none of the "distractions" which might have created alternative or supplementary sectors in the local economy.[50] The linkages of the staple, that is, either were weak or were suppressed to a considerable degree.

Comparison of the fishing areas around the Baie des Chaleurs with the nearby areas involved in other staples, although suggestive rather than definitive, reinforces this impression. In the Miramichi Basin, for example – which was settled anew in 1764 after the Conquest, at the same time that Charles Robin reached Paspébiac – timber was the predominant staple, supported by supplementary farming and fishing. Poor though it is said to have been, the Miramichi was significantly ahead of Gaspé in 1831 by any measure of economic expansion. The gross imports of the port of Miramichi for the years 1828–1830, for example, were £395,455 currency, while its exports were £400,136 currency, a favourable "balance of trade" of £4,681 currency.[51] Table 21 compares the Miramichi and Gaspé regions in terms of saw and grist milling per head of population.

Table 21
Sawmills and Gristmills per Head of Population, c. 1830

	Sawmills	Gristmills
Gaspé County	–	1:5003
Bonaventure County	1:1770	1:1061
Gloucester County	1:1083	1:812
Kent County	1:441	1:540
Northumberland County	1:513	1:711
Boiestown	1:60	1:120

Source: Census, 1831; R. Cooney, History, 1832.

Even in the hinterland lumber town of Boiestown, there were more gristmills than there were in Gaspé or Bonaventure counties, and Boiestown also boasted two sawmills, a forge, a washing mill, a school, and a hotel, all servicing a population of 120 persons. Northumberland and newly formed Kent County were likewise well ahead of Gaspé, and Cooney commented of them that the population was increasing, roads had been opened, bridges erected, schools founded, agriculture extended, cattle improved, fisheries were increasing, and several large vessels had been built. Even Gloucester County – bordering the south shore of the Baie des Chaleurs and directly across from Gaspé – which had the poorest performance on the New Brunswick north shore, was doing better, despite its small, localized Jersey cod fisheries in places such as Caraquet and Shippegan. Cooney, however, observed of this county that "we have made little progress. And why? because the water has been unexplored – the forest over-levied – and the soil neglected"; he advised that the solution was to "lumber moderately – Fish vigorously – and Farm steadily …"[52] Nonetheless, Gloucester County, although smaller in area than the Gaspé Peninsula, had a more diversified economy, with more gristmills, and more of them per head of population. Moreover, it is worth remembering that the economic development of Miramichi was far behind that in places like lakeside Ontario sixty years after initial settlement there.[53] Miramichi, however, is a more realistic basis for comparison with Gaspé since the areas are contiguous and were settled around the same period, the real difference between them being the staple on which each region was built. Miramichi performed very modestly in terms of economic growth, but Gaspé was much worse off. In Gaspé County, for example, only 0.1% of all households were involved in saw or grist milling (0.02% of the population), while in Bonaventure County

0.85% of all households were involved (0.15% of the total population)
– and sawmills and gristmills were the only manufacturing establishments that Gaspé had. What little growth there was remained
almost entirely related to the Jersey merchant fishery.

It is fair, then, to conclude that the resident merchant fishery did
little to expand or diversify the economic base of the region, but
there remains the question of whether the firm and its colonial merchant structure were entirely responsible for the lack of development
in the area or whether the nature of the staple also played a part in
this. In other staple trades, such as timber and wheat, development
did occur to a greater or lesser extent, depending on local conditions
and on factors affecting the wider economic picture. These other
staples, however, differed in one significant respect from the fishery,
and that was in the spatial diffuseness of the resource, or its inac-
cessibility, or both. In timber, for example, the staple had to be
sought at increasing distances from the coast, and this forced pen-
etration of the hinterland. In wheat, large tracts of land had to be
cleared and opened for agriculture, thereby again penetrating the
hinterland. In such resources as gold, oil, and even iron, some pen-
etration of the hinterland usually occurred, and the fur trade can
claim much of the credit for providing the motivation that first led
to the exploration of Canada's northwest. Such staples as fur, timber,
and wheat, also because of the spatial diffuseness of the resource,
had to evolve long chains of supply lines (in the fur trade)[54] or
control points such as those described by Wynn for the New Brunswick timber trade.[55]

In the fishery, however, this extension of control systems was
neither needed nor desired. Moreover, the firm maintained entrepreneurial control and management training within its own system,
and hence (by extension, if not by design) hindered the development
of a local entrepreneurial class. The organizational skills of the Jersey
merchants, regarded with admiration and envy by observers, were
a vital element in their control of the staple, and Fortin felt that this
was of primary importance in describing their success:

Rien de plus beau que l'ordre, la propreté et l'économie qui règnent dans
ces établissements. Aussi éxige-t-on des différents commis employés dans
le commerce du poisson un apprentissage régulier qui dure plusieurs an-
nées. Il n'y a pas un agent supérieur qui n'ait eu pendant longtemps la
charge d'un petit établissement, où il a du donner des preuves de son activité
et de sa capacité; pas un premier commis qui n'ait d'abord appris, en oc-
cupant des emplois inférieurs, à bien juger de la valeur de marchandises,
de la qualité du poisson.[56]

Such finely concentrated control of management skills could not readily have been achieved in a spatially diffuse staple trade such as timber, and the Jersey policy of maintaining Jerseymen as clerks is no more than a reflection of their ability to use the concentration of that particular staple industry to maintain a high degree of quality control in the selection of clerical staff without having to involve themselves in the more extensive, and therefore expensive, training that would have been needed had local persons been used. Baldwin, as well as pointing out the poor demand factors generated in the plantation economies of the Carolinas, noted that the supply factors of rate of saving and entrepreneurial labour in these economies contributed to the lack of development of domestic production.[57] In the case of a continued influx of cheap labour from Jersey into the Gaspé cod fishery, along with the colonial/mercantile control mechanisms that operated on the coast, the low-income factor remained constant, while savings generated by the economy were either reinvested in the staple production unit or else accrued to Jersey, in the form of importation of goods, salaries of Jersey labourers, and merchant profits.

Of course, the root problem with the fishery as a staple trade, from the merchant's point of view, was that control of access to the resource was seriously compromised by unprotected rent; therefore, it had to be tightly controlled and protected. Once the mechanisms for control were evolved, however, the industry achieved a concentration of power and control that was more intensive than that created in any other nineteenth-century staple. Just as the fishing village was the industry, so – by extension of function – was the merchant. He became at once supplier, importer, distributor, producer, processor, collector, exporter, marketer, and financier of the Gaspé fishery.

The spatial concentration of the fishery on a littoral, combined with the functional integration that almost automatically results, can be regarded as inherent characteristics of the staple itself. The causes of inadequate development in Gaspé become clearer when they are considered within the context of the exogenous nature of the nineteenth-century merchant triangle along with its colonial financial and control structures (such as the truck system and the import/export monopoly). Jersey merchants were obviously less interested in any potential that Gaspé might have developed as a consumer market than they were in its functions as an access base for the fishery: CRC's interest was focused on cheap fishing and cheap labour and its ability to supply and control both, maintaining its monopsony in the process. Indeed, there was minimal interest in the area by anyone, other than that very specific interest shown by the Jersey merchants. Even the colonial government, which controlled the area politically, dismissed any consideration of government development of its fisheries, despite arguments that such interest would go a long way toward breaking the Jersey monopsony in the area.[58] The net result was that what little capital existed tended to flow out of the area, back to the mother country, as payment for the few luxury imports that were demanded, while no counterbalancing flow accrued to Gaspé. Indeed, any savings that were generated in the local economy clearly had to go back into the fishery as re-investment in the production unit. In other words, so long as the truck system and the import/export monopoly restricted the cash economy and curtailed earnings, the low-income factor was held constant and final demand remained negligible.

The fishing economy in Gaspé, then, can be said to have generated growth, but little or no development as defined at the beginning of this chapter. How much diversification might realistically have been generated in the area is, of course, a matter for conjecture: it is impossible to estimate with any degree of certainty the potential lost by virtue of the way in which the merchant inshore fishery was operated. But it is clear that mercantile restriction of linkage formation in Gaspé inhibited development and contributed significantly to the economic retardation of the region. This is not to say, however, that the economic benefits (such as they were) of the fish staple were entirely lost, but to observe that very few of them were found in Gaspé.

CHAPTER SIX

Consequences of the Trade: Jersey

The history of the cod trade in Gaspé is an account of staple-trade growth, but not of regional development. If, however, the economic effects of that trade on Jersey are examined, a very different picture emerges. In Jersey, the cod trade of Gaspé was seen as providing an economic base for the island which it did not possess *in situ*. By 1837, Jerseymen were commenting that the codfishery was of prime importance to Jersey, not only because of the value of the industry *per se* in terms of labour employed, capital invested, and the returns on that investment, but also because it was "the root of other indirect industry, and the means of supporting many families."[1] Indeed, it was thought that "without her codfishery, the commerce of Jersey would dwindle away."[2] In other words, the claim was being made that the cod trade was providing the kind of economic leading edge for Jersey that is to be expected of a successful export staple trade: the linkage effects that were absent in Gaspé were being identified as benefits for the metropolitan end of the trade.

Looking at Jersey in 1840, when the cod trade was reaching maturity, it is certainly clear that the island was entering a period of growth and prosperity. Jersey's fisheries were being prosecuted in British North America on an equal footing with the United Kingdom's, and the mercantile community was seeking to extend its commercial interests on a broader front.[3] In 1841, Philip De-Quetteville, president of the Chamber of Commerce, wrote to Lloyds seeking A–1 classification for Jersey vessels, built – as he pointed out – specifically for merchants on their account, and not on speculation.[4] Such A–1 classification was important for an island which was becoming increasingly involved in the carrying trades and which was therefore dependent on charter voyages and hence on the status of A–1 ships, not to mention on the increasing need for insurance at favourable rates. DeQuetteville pointed out that "at the United States our vessels, not of the first letter, can be insured on more favourable terms than A–1 English ships,"[5] and he stressed the increasingly good reputation of the Jersey fleet in world trading centres. The records of the Custom House in London likewise give ample evidence of the successful expansion of Jersey into a wider commercial sphere. Also in 1841, for example, Philip Pellier, a fish merchant, wrote the Customs as follows:

Gentlemen:

As I am on the eve of sending out my new Barque the 'Achilles' of 288 tons (N.M.) to the Cape of Good Hope and the Mauritius – and several parties with myself being desirous of furnishing goods on Speculation to those Colonies, among which I would enumerate "Bricks, Coals, Salt, Cordage, Soap, Vinegar, Flour, Biscuit, Iron, Coal Tar, Ironware, Beer, Piece Goods, Mus[l] Instruments, Stationery", *British Produce or Manufacture* – "Wine, Geneva, Deals, Spars, Pitch, Tar, etc." – *Foreign* – and a direct intercourse between these Places being a new feature in the Trade of this Island, May I request ... under what footing Such Trade may be carried on ...

I am, Gentlemen, etc. ...
Ph. Pellier (Owner).[6]

In 1845, the Chamber of Commerce noted that "now vessels from the Island frequent all parts of the Globe. You will find them in the South Seas at our Antipodes. They may be seen at New Holland and in the Indian Ocean."[7] Shipping entering inwards to Jersey with cargo had risen from 1,360 vessels, in 1838, to 1,585 vessels, by 1844, and shipping registered at Jersey had increased from 244 vessels (23,826 tons, old measurement) to 311 vessels (28,078 tons) over the same period.[8] The Chamber of Commerce suggested that St Helier needed a new dock, and shortly thereafter (in 1851) the island acquired its own local board of examiners for the mercantile marine.[9]

Ancillary communication services were also expanding. By 1846, the mail service to Jersey was being improved, and in 1854, with trade and navigation still on the increase, the Chamber sought to have the island mail service put on a daily footing.[10] In 1858, telegraph service to the island was inaugurated.[11] By this time there were 372 vessels (33,000 tons) registered in Jersey, and the fisheries were also growing: in 1853 they had directly employed about one hundred vessels, with a total of over 10,000 tons, along with about two thousand seamen.[12] By 1860 there were 60,000 tons of shipping in the whole of the Channel Islands – a growth rate of 57% was achieved between 1831 and 1863[13] – and the Chamber of Commerce thought that completely new harbour facilities, capable of handling both sail and steam vessels, might be necessary.[14]

In line with the general expansion of the island economy, the population also grew. It had risen from the 25,000 inhabitants of the early 1600s to 60,000 persons by 1860,[15] only 15,000 fewer than present-day figures. The annual rate of increase of population between 1824 and 1851 was 2.23%, compared to 1.6% for Guernsey and 0.88% for the Isle of Man, both of comparable size to Jersey. The rate for England and Wales over the same period was 1.28%, and that for Scotland was 1.04%, although the considerable regional variations that existed in these latter areas make direct comparison hazardous.[16] There was no sign, either, that large-scale emigration (the perennial fear of the Chamber of Commerce) might occur.[17] Jersey was experiencing major population growth without it being in any way a distressing increase which might overburden the economic resources of the island.

Clearly, between 1830 and 1860 the economy of the island had expanded very rapidly. The question is whether or not this was in any way linked to the success of the cod trade. If the two phenomena were related, there should be evidence that Jersey had concentrated initially on the security and well-being of the fish trade – and that has already been established in chapters 1 and 3. There should also have been an expansion of cod-merchant capital into related (that is, structurally linked) areas of the Jersey economy such as shipbuilding. Thereafter, a more general broadening of the economic base could be expected. Cod merchants would not have had to be dominant in the later stages of expansion, but their initial input into growth sectors would have had to be early and substantial. In other words, the forward, backward, and final-demand linkages of the cod trade should have been evident as being the principal engines of growth in the indisputable economic development of the island economy in these years.

The anonymous author of an 1837 article on Jersey had no doubt that the island's burgeoning economy was an outgrowth of the cod trade. He stated categorically that "by far the most important and beneficial branch of the commerce of Jersey, are the fisheries on the coasts of British North America. That branch is not only valuable from the direct industry which it promotes, the capital which it employs, and the number of persons who are engaged at the fishery, but it is the root of other indirect industry, and the means of supporting many families. Without her codfishery, the commerce of Jersey would dwindle away."[18] He justified his statement by explaining that "at the present time a large capital, a great number of vessels, and many persons are engaged in the fisheries"[19] – 1,275 Jerseymen were employed therein[20] – and then went on to offer a detailed picture of what were in fact linkages captured for Jersey from the cod fishery. Tantalizing though this contemporary claim is, however, it does not substantiate, so much as assert, the role of the cod trade in Jersey's development. To prove it, there would have to be concrete evidence of the connections between growth sectors of the island economy and the cod trade. That is the purpose of this chapter.

Forward linkages from the codfishery, as previously noted, were very weak, and were in fact found (so far as the technology of the time permitted) in Gaspé, with the dry cure. The cured fish was consumed in the market countries as processed on the Gaspé coast; no further processing occurred at the markets, or in Jersey.

The major potential backward linkage from the fishery was shipbuilding, which was carried out for some time in Gaspé, but after 1834, even CRC's shipbuilding in Gaspé was minimal. Figures 19A and 19B show that shipbuilding by Jerseymen occurred initially primarily in the New World and remained most important there up until about 1819, at which time the trend toward Jersey-based shipbuilding became more pronounced; by 1834 it predominated.[21] Indeed, in 1850 CRC was to write to Canadian lumber merchants in Campbelltown, "We beg to state that, having discontinued shipbuilding, we are not in want of juniper, or indeed any other kind of timber."[22]

There was considerable capital investment in the vessels built in Jersey in the middle years of the nineteenth century. The Chamber of Commerce noted in 1845 that "it may be that the increase hereafter may not be so proportionately great or so rapid as during the past, but with increasing capital and enterprise we may reasonably believe that an increase of the shipping and commerce of the Island will continue."[23] The question, however, is: Who was investing in this shipbuilding and was it connected to the cod trade?

Ship registers provide ownership information but do not provide information on voyages; however, the Jersey newspaper *Chronique de Jersey* published shipping news ("Nouvelles de Mer") once a week in 1830 and twice-weekly by 1840. The "Nouvelles de Mer" gave arrivals and departures for Jersey vessels at all ports at which they called, and provided data that can be reconstructed to show the

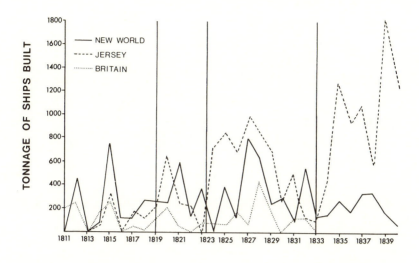

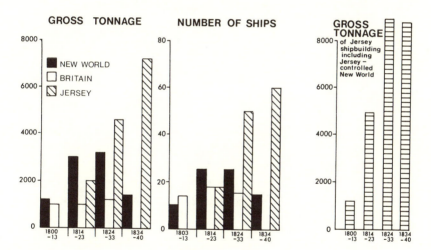

Figure 19. Tonnage by Place of Construction

network of shipping that developed in the 1830s and 1840s. One example will suffice to demonstrate the technique. Figure 20A shows the voyage of one vessel which was, in fact, a CRC vessel; all voyages for 1830 and 1840 were reconstructed in this way, and gaps in the data were filled by an estimation of the latest possible date at which the missing section could end. (Thus it was necessary for the *Teaser* to be in Jersey by April if she were to arrive in Gaspé by early May.) Figure 20B shows the next step in data sorting. Ports of call existed

A.

7-12-1839	Baie des Chaleurs	to	Valencia	in
29-12-1839	Valencia	to	Cadiz	out
1-1-1840	Valencia	to	Cadiz	in
circa 4-1840	Cadiz	to	Jersey	in
12-5-1840	Jersey	to	Gaspé	in
circa 7-1840	Gaspé	to	Jersey	in
31-8-1840				
or 1-9-1840	Jersey	to	Cadiz	in
26-9-1840	Cadiz	to	Alicante	out
5-10-1840	Cadiz	to	Alicante	in
22-10-1840	Cadiz	to	Messina	in
6-11-1840	Messina	to	Palermo and London	out
9-11-1840	Messina	to	Palermo	in
23-11-1840	Palermo	to	Falmouth	out

B.

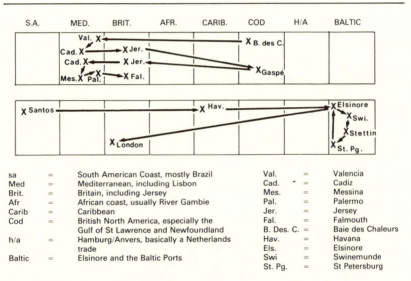

sa	=	South American Coast, mostly Brazil	Val.	=	Valencia
Med	=	Mediterranean, including Lisbon	Cad.	=	Cadiz
Brit.	=	Britain, including Jersey	Mes.	=	Messina
Afr	=	African coast, usually River Gambie	Pal.	=	Palermo
Carib	=	Caribbean	Jer.	=	Jersey
Cod	=	British North America, especially the	Fal.	=	Falmouth
		Gulf of St Lawrence and Newfoundland	B. Des. C.	=	Baie des Chaleurs
h/a	=	Hamburg/Anvers, basically a Netherlands	Hav.	=	Havana
		trade	Els.	=	Elsinore
Baltic	=	Elsinore and the Baltic Ports	Swi	=	Swinemunde
			St. Pg.	=	St Petersburg

Figure 20. Voyages of the *Teaser*

in various areas which could be broadly classified according to commodity groups, and this classification was used to group ports together in order to chart voyages in a consistent and comparable
manner. Names of ports were retained, however, in case separate
trades were found to occur within a commodity area.[24] Even ports
which, strictly speaking, were outside the geographical limits of an
area were included if it was thought that their principal commodities
were part of the nearby regional commodity group. Thus Lisbon,
trading in salt, fruit, and wine, was included in the Mediterranean
group. No attempt was made, at this stage, to identify specific voyages as belonging to one or more trade networks.

With the data in this form, it was possible next to discover which were the most heavily used trade routes and the connections between one trade area and another. Short coastal voyages were not included, since the long-distance trades were the new investment growth area for this period. Taking the passage from one port of call to the next as one "leg" of a voyage, the numbers of legs between trade areas were calculated. This gave an initial impression of the volume of traffic and of its direction. Then the exercise was repeated for subsequent legs or trips, thus indicating where, given arrival at one destination, a vessel was most likely to go next. These second legs are shown in tabular form in tables 22 and 23 in the "second order" columns. They are read as follows: In 1830, given that eleven vessels voyaged from South America (SA) to Britain, on the next leg of the voyage one could be expected to go back to South America, one to another place in Britain, and four to Hamburg/Anvers. The rest would have terminated their voyages. That is, after reading the first-order matrix for South America to Britain, the next leg is picked up by reading along the row SA/Br. The likelihood of a further leg can be assumed, but – because the matrices were constructed on a non-voyage basis – all legs were counted once as "first order," then their subsequent legs were entered on the second-order matrix; extension of the technique to further legs is possible, but quickly becomes very unwieldly and the numbers very small.

This is a useful way of dealing with a one-year sample of voyage data, because it shows connectivities between places even when complete voyages cannot be captured by the one-year sample, as is often the case. Nor could the sample realistically be extended given the constraints of this study, although an extended study of this kind, concentrating on Jersey itself, would obviously be a rewarding exercise. Instead, what has been done here is to include from time to time complete voyages where such information clarifies the discussion. No purely local coasting voyages have been included in the analysis: where coasting legs are shown they are always a coasting component of longer-distance voyages. The result of this analysis was the identification of a variety of trading networks or geometries, the connections between them, and how they changed between 1830 and 1840.[25]

Figures 21 and 22 show the networks for 1830 and 1840 for all connections of five trips or greater. In reading these figures and the subsequent two derived from them (figures 23 and 24), two inherent biases must be borne in mind, both of which arise from the limitations of the data. The first is the built-in overemphasis on short hauls, such as the two-way Mediterranean trades. Obviously, more

Table 22
Trade Area Connectivity, 1830

	SA	Med.	Brit.	Carib.	Cod	HA	Baltic	Totals
				First Order				
SA	13	6	11	–	–	14	1	45
Med.	17	25	40	1	5	2	1	91
Brit.	12	62	23	11	30	6	6	150
Carib.	–	1	5	2	7	1	–	16
Cod	4	21	8	7	8	–	–	48
HA	–	2	9	–	–	1	1	13
Baltic	–	1	7	–	–	–	3	11
Total	46	118	103	21	50	24	12	374
				Second Order				
SA/SA	3	2	2	–	–	4	–	11
SA/Brit.	1	–	1	–	–	4	–	6
SA/HA	–	1	4	–	–	1	–	6
Med./SA	6	1	3	–	–	4	–	14
Med./Med.	4	3	11	–	1	–	–	19
Med./Brit.	1	3	8	1	2	–	–	15
Brit./SA	3	1	3	–	–	2	1	10
Brit./Med.	2	14	15	–	3	–	1	35
Brit./Brit.	–	1	3	–	2	1	–	7
Brit./Carib.	–	1	4	1	1	1	–	8
Brit./Cod	2	5	3	2	7	–	–	19
Cod/Med.	–	6	6	–	1	–	–	13
Cod/Brit.	–	2	1	–	2	–	–	5
Cod/Carib.	–	–	–	–	7	–	–	7
Cod/Cod	1	5	1	–	2	–	–	9
Total	23	45	65	4	28	17	2	184

Source: "Nouvelles de Mer," *Chronique de Jersey*, 1830, passim.
Key: SA = South America; Med. = Mediterranean; Brit. = Britain, including Jersey; Carib. = Caribbean; Cod = British North American fisheries; HA = Hamburg/Anvers.

of these could occur in a given year than could long-haul transatlantic trades. Second, in any long-distance trade in which there was a strong seasonal component which emphasized the very early, or very late, months of the year, early or final stages of that trade were likely to be lost. For example, if vessels from South America arrived at a Mediterranean port en route for Jersey in mid-December, the final passage to Jersey would in all likelihood be absent from the diagrams. Nothing can be done about either of these problems

Table 23
Trade Area Connectivity, 1840

	SA	Med.	Brit.	Afr.	Carib.	Cod	HA	Baltic	Totals
				First Order					
SA	7	6	26	3	2	2	9	–	55
Med.	17	75	91	–	1	22	–	1	207
Brit.	30	94	92	8	22	41	19	29	335
Afr.	4	–	6	–	–	–	1	–	11
Carib.	–	2	20	–	5	1	1	1	30
Cod	7	32	19	–	3	17	–	–	78
HA	–	–	18	–	1	1	6	3	29
Baltic	–	–	46	–	–	1	1	48	96
Total	65	209	318	11	34	85	37	82	841
				Second Order					
SA/Brit.	–	–	7	–	–	–	10	–	17
SA/HA	1	–	1	–	1	1	2	2	8
Med./SA	2	–	6	1	–	–	1	–	10
Med./Med.	1	29	26	–	–	5	–	–	61
Med./Brit.	–	3	30	–	–	9	–	–	42
Med./Cod	2	8	3	–	–	3	–	–	16
Brit./SA	2	3	10	1	–	–	1	–	17
Brit./Med.	11	39	14	–	–	10	–	–	74
Brit./Brit.	3	6	24	3	3	2	1	1	43
Brit./Carib.	–	–	13	–	5	–	–	–	18
Brit./Cod	–	13	7	–	1	11	–	–	32
Brit./HA	–	–	11	–	–	–	1	1	13
Brit./Baltic	–	–	2	–	–	–	–	22	24
Carib./Brit.	–	2	8	–	–	–	1	–	11
Cod/Med.	–	23	4	–	–	3	–	–	30
Cod/Brit.	–	7	4	–	–	1	–	–	12
HA/Brit.	3	1	5	–	–	–	1	–	10
Baltic/Baltic	–	–	26	–	–	1	1	15	43

Source: "Nouvelles de Mer," *Chronique de Jersey*, 1840, passim.
Key: SA = South America; Med. = Mediterranean; Brit. = Britain, including Jersey; Carib. =
Caribbean; Cod = British North American fisheries; HA = Hamburg/Anvers; Afr. = Africa.

within the given analytical constraints except to recognize that they
exist.

Looking first at 1830 (figure 21), the pattern was one of relatively
discrete older trading networks along with a few newcomers. As
would be expected with an important short haul, the Jersey (Britain)
link with the Mediterranean was dominant. Next in importance was
the link to the cod trade. As with the single example of the *Teaser*

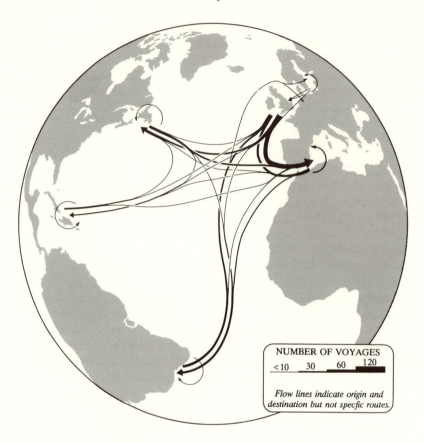

Figure 21. Jersey Voyages, 1830

(figure 20), these two trade areas were related, the Mediterranean being the traditional market area for the cod trade. There was also a simple two-way Jersey-to-Mediterranean-to-Jersey trade, essentially an extension of the coasting trade that had existed for a long time. Typical of this latter trade was the voyage of the *Twig* in 1830, going from Jersey to Malta to Messina to Jersey and back again to Malta. The two different Mediterranean trades (coastal and cod) cannot be distinguished in the figure.

The 1830 figure also shows passages to the Caribbean and to South America, the former closely connected to the cod trade on a first-order basis, the latter slightly so if second-order links are considered. The Caribbean trade was the old West Indies trade; the South American trade was new – these markets opened up only after Brazilian independence in 1822. There were also a few voyages to the Baltic

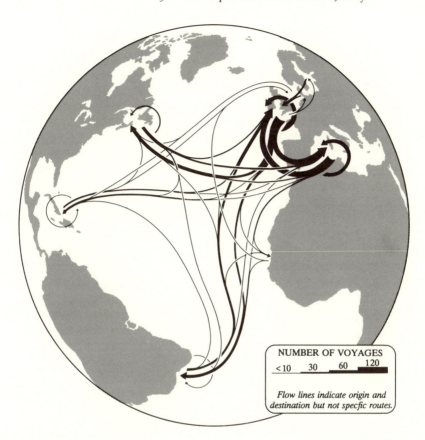

NUMBER OF VOYAGES

<10   30   60   120

*Flow lines indicate origin and
destination but not specfic routes.*

Figure 22. Jersey Voyages, 1840

and to Hamburg/Anvers, with the latter also connected to South
America. Such was the voyage of the *Phoenix*, which sailed from
Jersey to Bahia to Anvers to Jersey to Rio to Anvers in 1830. The
most important points to take from the 1830 figure, however, are
the multiple connectivities of the cod trade, second only to those of
Jersey (Britain) itself, and the connectivities of South America. Together, these two New World trades show a considerable potential
for trade diversification.

By 1840 (figure 22), the network was much bigger and more complex, and some shifts had occurred. First, the collapse of the
Caribbean-cod connection is obvious – a result of the emancipation
of slaves in 1834 and the consequent destruction of the slave-plan-
tation market for cheap, poor-quality codfish.[26] The Caribbean was,
in fact, being replaced by South America. There was also a new

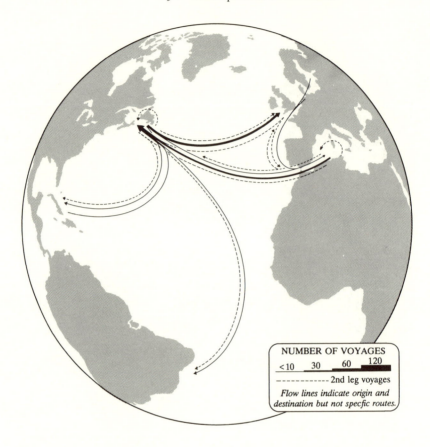

NUMBER OF VOYAGES

<10   30   60   120

- - - - - - - - - - 2nd leg voyages

*Flow lines indicate origin and
destination but not specfic routes.*

Figure 23. The Cod Trade, 1830

African trade, mostly in ivory and other such products, which also
arose out of the abolition of slavery and a consequent shift in the
African trades. Last, there was a considerable strengthening of the
Baltic trades, along with a slight weakening of the South America-
to-Hamburg/Anvers connection. The reasons for the rise of the Baltic
will be dealt with later. In the meantime, the major points to be
drawn from this figure are the impression of overall growth, the
continued strengthening of the connectivities of the cod trade, and
the stronger inclusion of South America in the picture, along with
a general impression of greater connectivity for the whole network.

With that as a backdrop, the role of the cod trade in Jersey's trading
system can now be further examined. Figures 23 and 24 show the
first- and second-order links of the cod trade for 1830 and 1840,
exhibiting diagrammatically something of how links between pas-

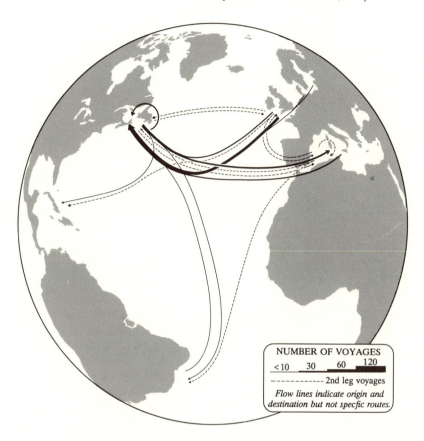

NUMBER OF VOYAGES
< 10   30   60   120
------------ 2nd leg voyages
*Flow lines indicate origin and
destination but not specfic routes.*

Figure 24. The Cod Trade, 1840

sages occurred. The same *caveat* about seasonality applies here as
in figures 21 and 22, but the distance bias is much reduced by the
elimination of really short hauls. What the 1830 figure shows is the
pattern that is generally accepted as the classic cod-trade picture:
supplies *in* from Jersey and Britain, fish *out* to markets in the Caribbean, South America, and the Mediterranean. In this figure the
connectivities to South America show up, thereby highlighting the
strength of the methodology at this level. Clearly, the more legs that
can be included diagrammatically the better; however, the cartographic problems of so doing are considerable.

By 1840 (figure 24), the cod trade looked rather different from the
traditional picture. It was larger, as one would expect. It was also
much more complex and much more integrated into the total system.
This figure shows the cod trade as viewed from the metropole, and

it should be compared with figure 10, the view from the periphery. Even Hamburg and the Baltic were connected to the cod trade by this time: the *Guernsey Lily* went from Rio de Janeiro to Hamburg to Dantzig to Jersey to Cadiz to Gaspé; the *Venus* went from New-foundland to Bahia to Hamburg to Dantzig to Jersey to Cadiz to Newfoundland. It is clear from voyage information alone that the cod trade was important to the Jersey network. It was complex, with multiple and growing connectivities, and was the most important of the Jersey transatlantic trading geometries.[27] Nonetheless, while the voyage analysis is therefore supportive of the 1837 statement that the cod trade was "by far the most important and beneficial branch of the commerce of Jersey," it does not completely confirm that claim. To do that, substantial cod-trade ownership of vessels in the early days of the rise of Jersey shipping has to be demon-strated. Unfortunately, voyage analysis is not possible for the earliest years of the ship registers, because the "Nouvelles de Mer" were not being published then. However, by 1830 they contained 92% of all Jersey shipping on the registers. By 1820, when Jersey-built ton-nage overtook tonnage built elsewhere, it was dominated by cod-trade merchants: from 1818 to 1821, tonnage owned by cod mer-chants was always more than 90% of all new Jersey-built vessels, and from 1822 to 25 it was still more than 50%. Cod-trade merchants built large vessels, while an increasing number of small vessels, usually coasters, were built by other owners.

By 1830, the ship-register information can be combined with the "Nouvelles de Mer" to produce a more fine-grained analysis which identifies owners of vessels on specific routes. Table 24 shows the distribution of ownership by vessel and by tonnage among the var-ious trades of Jersey, using only those vessels from the ship registers which were also included in the "Nouvelles de Mer," since it was only for those vessels that voyage information was available.[28] In 1830, the cod trade predominated in number of vessels, tonnage, number of vessels per owner, and tonnage per owner. The coasting trade predominated in number of owners, but the tonnage per vessel in this trade was very low, and the trade consisted entirely of one-vessel owners – a pattern of local, small-scale enterprise, often run by mariners or small commercial entrepreneurs such as butchers and innkeepers. In contrast, the cod trade was composed of a much higher average tonnage per owner, indicating a much greater in-vestment in the trade. As well, in the cod trade was found the merchant-owner with by far the greatest number of vessels. Clearly, in 1830 investment was both higher and more concentrated in this trade than in any other.

Table 24
Owners and Trades, 1830

Trade[1]	Number of Vessels	Percentage of Total Shipping	Number of Owners	Tonnage	Average number of Vessels per Owner	Most Vessels per Owner	Tonnage per Owner
Cod	44	27	20	5,694	1.5	9	285
Med./Brit./Med.	33	20	25	3,142	1.3	3	126
Coasting	30	19	30	1,355	1	1	45
SA to HA	16	10	13	2,183	1.2	3	168
Caribbean	14	9	11	1,913	1.2	3	174
Baltic	10	6	8	1,515	1.25	2	189
SA to Brit.	6	4	6	1,823	1	–	172
Others	8	5	6	877	–	3	146
Total	161	100	119				

Source: Jersey Ship Registers
Note: [1] Vessels were allotted to only one trade per voyage.

Table 25
Index of Firm Diversification, 1830

No. of Firms	Trade								Total	Rank
	1	2	3	4	5	6	7	8		
1 19	–	.37	.16	.26	.37	.16	.05	.26	1.63	1
2 19	.37	–	.11	.16	.05	–	.05	.16	0.9	4
3 19	.16	.11	–	.05	.11	.05	.05	–	0.53	6
4 14	.36	.21	.07	–	.07	.14	–	.07	0.92	3
5 11	.64	.09	.18	.09	–	–	–	–	1.00	2
6 9	.33	–	.1	.2	–	–	–	–	0.63	5
7 3	.33	.33	.33	–	–	–	–	–	1.00	2
8 10	.5	.3	–	.1	–	–	–	–	0.9	4

Note: 1 = Cod   5 = Caribbean
      2 = Med   6 = sa to Med/Brit
      3 = Coast   7 = Brit/Med to Baltic
      4 = sa to ha or Baltic 8 = Other

Cells computed as: Of the 19 firms in the cod trade, .37 (37%) were also involved in the
Mediterranean trade, etc.

Moreover, merchant firms diversified into various "international"
trades, as table 25 shows.[29] In this matrix, zero (o) represents the
highest possible degree of firm concentration in one trade (*no over-
lap*). That is, all firms in trade X are in trade X only, and therefore
trade X has a diversification index of zero. Conversely, then, the
highest index figures represent the greatest amount of diversification
by a firm into other trades. Rows can (and do) sum to greater than
1.00, since firms could (and did) diversify into more than one trade.
Thus for row 1, column 2, the value is derived as follows: of the
nineteen firms in the cod trade (trade #1), .37 (37%) were also involved in the Mediterranean trade (trade #2). Because of the small-
ness of the numbers involved, little more can be said than that the
cod-trade firms were most highly diversified, that the Caribbean,
Baltic, South America-to-Hamburg/Anvers, and Mediterranean
trade firms were moderately diversified, and that the South America-
to-Mediterranean-to-Britain and the coasting trade firms were the
least diversified. That is, there was a greater tendency for people
involved in the cod trade to diversify into other trades than there
was for any other commodity trade group.

Table 26 shows the distribution of ownership by vessel and by
tonnage for 1840. The enormous expansion of the coasting trade is
evident here: it had by far the most tonnage overall, and the highest
number of vessels. Tonnage per owner had also increased over the

Table 26
Owners and Trades, 1840

Trade	Number of Vessels	Tonnage	Percentage of total shipping	Number of Owners	Vessels per owner	Tonnage per owner	Most vessels per owner
Cod	56	7,051	14.7	31	1.8	227	8
Med./Brit./Med.	42	3,938	11.1	36	1.2	109	4
Coasting	131	8,253	34.5	106	1.2	78	6
sa/ha or Baltic	20	2,756	5.3	17	1.2	162	4
Caribbean	22	3,893	5.8	17	1.3	229	5
ha and Baltic to Brit./Med.	56	4,202[1]	14.7	52	1.1	81	–
sa to Med. to Brit.	31	4,730	8.1	29	1.1	163	3
Black Sea	6	938	1.6	6	1	156	–
Other	16	–	4.2	13	–	–	–
Total	380		100	307			

Note: [1] Estimated, based on average of vessels whose tonnage was known.

Table 27
Index of Firm Diversification, 1840

	No. of Firms	Trade									Total	Rank
		1	2	3	4	5	6	7	8	9		
1	26	–	.31	.62	.35	.27	.27	.12	.04	.15	2.13	1
2	30	.27	–	.17	.07	.07	.27	.13	.07	.03	1.08	2
3	51	.31	.1	–	–	.02	.08	.06	.1	.04	0.71	9
4	14	.64	.14	–	–	.21	.07	–	–	–	1.06	3
5	17	.41	.12	.06	.18	–	–	–	.06	.06	0.89	5
6	24	.29	.33	.17	.04	–	–	–	–	.04	0.87	6
7	12	.25	.33	.25	–	–	–	–	–	–	0.83	7
8	5	.2	.4	.2	–	.2	–	–	–	–	1.00	4
9	12	.33	.08	.16	–	.08	.08	–	–	–	0.73	8

Note: 1 = Cod Trade   6 = SA to Med/Brit
     2 = Med   7 = Brit/Med to Baltic
     3 = Coast   8 = Black Sea
     4 = SA to HA or Baltic   9 = Other
     5 = Caribbean
Cells computed as: Of the 26 firms in the cod trade, .31 (31%) were also involved in the
Mediterranean trade, etc.

decade, from forty-five tons to seventy-eight tons. The Baltic trade
was larger: by 1840 it had the same number of vessels as the cod
trade. However, many of these vessels were probably not Jersey-
owned,[30] and it might therefore be argued that the fifty-six vessels
of the cod trade were more important to Jersey than were the fifty-
six vessels of the Baltic trade, at least in terms of capital investment.
Of the international trades, tonnage remained highest in the cod
trade, and tonnage per owner was highest jointly in the cod and
Caribbean trades (227 tons per owner in the cod trade, 229 tons per
owner in the Caribbean trade). The latter, however, had less than
half the number of vessels (twenty-two, as opposed to fifty-six) and
a smaller number of vessels per owner (1.3, as opposed to 1.8). In
1840, then, the cod trade remained dominant in terms of number
of vessels per owner and firms with the largest number of vessels;
investment was still highest and most concentrated there. Table 27
shows the Index of Firm Diversification for 1840; once again, the
cod-trade firms were more diversified than those of any other trade,
now to an increasing degree (1.63 for 1830, up to 2.13 by 1840). The
coastal trade remained very little diversified, and even the Baltic was
no more than moderately so.

Tables 28 and 29 show the deployment of vessels of all firms listed
in tables 25 and 27 for 1830 and 1840, respectively. In 1830, the most

Table 28
Deployment of Vessels by Firms, 1830 (by Number of Voyages)

Firm	Cod	Med.	Coasting	SA/HA	Carib.	Baltic	SA/Brit.	Other
				Trade				
CRC	9	–	–	1	–	–	–	–
Nicolle	8	–	–	1	1	2	–	2
DeCarteret & LeVesconte	3	–	–	–	2	–	–	–
Duval	3	–	–	–	–	–	–	–
Janvrin	3	–	–	2	1	–	1	1
PRC	2	–	–	–	–	–	–	–
DeQuetteville	2	3	1	–	–	–	1	–
Renouf	2	–	–	–	–	–	1	–
Bisson	1	2	–	–	2	–	–	1
Deslandes	1	–	–	–	1	–	–	1
J. & P. LeBas	1	–	–	1	1	–	–	–
Ennis	1	2	–	–	–	–	–	–
Ranwell	1	–	1	–	1	–	–	–
Pellier	1	–	–	–	1	–	–	–
Roissier, Hamon & LeGros	1	1	–	–	–	–	–	–
Fruing	1	–	–	–	–	–	–	–
De Gruchy	1	–	–	–	–	–	–	–
Perchard	1	1	–	1	–	–	–	–
Bertram	1	1	–	–	–	–	–	–
Totals	43	10	2	6	9	2	3	5

diversified firm was Nicolle et Cie, based mainly in Newfoundland and trading into South America-to-Hamburg/Anvers, into the British-Baltic trade, into the Caribbean, into the Black Sea, and into the United States. Next was the firm of P. and F. Janvrin, which had its major cod establishments at Arichat, Cape Breton. This firm was also involved in the South America-to-Hamburg trade, the Caribbean trade, and the South America-to-Britain trade.[31] It also had one ship trading on the French coast (Nantes and Rouen). Then came the firms of DeQuetteville and Bisson, both of which were involved in three other trades. In terms of areas most used by cod-trade firms, the traditional trade areas of the Mediterranean and the Caribbean were most popular.[32] The Gaspé firms, however, especially CRC, seemed less inclined to diversify, and the question may be asked whether an expanding Gulf cod trade meant less need to diversify than might have been felt in Arichat and Newfoundland, where the older, more established firms were both more capable of diversification and perhaps more pressured into doing so.

Table 29
Deployment of Vessels by Firms, 1840 (by Number of Voyages)

Firm	Trade								
	Cod	Med.	Coasting	SA/HA	Carib.	Med./Brit./SA	Brit./Baltic/Med.	Black Sea	Other
Nicolle	8	–	1	1	1	–	–	–	–
CRC	7	2	–	–	–	–	–	–	–
DeQuetteville	6	4	2	–	–	–	1	–	1
Janvrin	3	–	1	–	1	1	–	–	–
PRC	2	–	–	–	–	–	–	–	–
Fruing	2	–	2	–	–	–	–	–	–
Perrée	2	–	1	–	–	–	–	–	–
Bréa	2	2	5	–	–	–	1	–	–
Decarteret & LeVesconte	2	–	2	1	–	–	–	–	–
LeFevre	2	–	1	–	–	1	–	–	–
Martell/Vibert	1	–	1	1	–	–	–	–	–
LeBoutillier	1	–	–	–	–	–	–	–	–
Amiraux, Marrett, LeBas	1	–	1	–	1	–	–	–	–
Woolcocke	1	–	–	–	–	–	–	–	–
Gossett	1	–	–	3	2	–	–	–	–
Fauvel/Godeaux	1	1	–	–	–	–	–	–	1
Messervy	1	1	–	1	–	1	–	–	–
Carrell	1	1	2	–	–	–	–	–	2
Renouf	1	–	–	–	–	–	2	–	1
Bayfield & Copp	1	–	1	1	–	–	–	–	–
E. Lebas	1	–	–	1	–	–	–	–	–
LeQuesne	1	1	2	1	–	1	–	–	–
Deslandes	1	1	2	1	–	1	–	1	–
Perchard	1	–	2	–	1	1	–	–	–
Roissier, Hamon, LeGros	1	–	1	–	–	–	–	–	–
Godfray	1	–	–	–	–	–	–	–	–
Totals	52	13	27	11	6	6	4	1	5

In 1840 (table 29), diversification in the cod trade had increased. DeQuetteville had added a vessel trading into Archangel, but DeCarteret and LeVisconte were no longer in the Caribbean – further corroboration of the collapse of the cod/Caribbean connection. Some of the apparent new growth in the cod trade was the result of one-time agents for the traditional large firms having set up in business on their own. Fruing and Fauvel are cases in point: they were off-shoots of CRC.[33] This pattern of "hiving-off" was a continuing phe-nomenon in the cod trade: J. and E. Collas, for example, started out as agents for Perrée, took over that firm in the 1850s, and finally took over CRC when the Robin family went bankrupt in 1886. Also worth noting is the firm of George Deslandes and Son, which was very small in 1830, but by 1840 was already the most diversified of all firms shipping into the cod trade. This firm was destined to become the largest of the Jersey ship-building firms of the 1840–60 "boom" era.

In summary, then, the cod trade was demonstrably extremely important to Jersey shipping. Owners in the cod trade commanded more, and heavier, vessels than those in any other trade, and they did so earlier. They were also, in the early days, more diversified, linking these separate trading networks into one complex structure. One explanation for this linkage can be found in the 1833 *Report* on the Gaspé district, which stated that Messrs Robin and Com-pany "have extensive commercial establishments in Brazil, Foreign Europe and other ports. They export their fish in their own vessels and bring return cargoes to Hamburg and other ports in foreign Europe."[34]

Table 30 shows the new registries at Jersey for the years 1834–39, the dominance of the Jersey-built vessels, and their larger average tonnage. The concomitant decline in New World-built tonnage is also obvious, further evidence that cod-trade shipbuilding in the New World was increasingly shrinking to the provision of shallops and non-ocean-going boats, leaving larger construction to Jersey. One member of the Robin family (C.W. Robin) registered his first Jersey-built vessel during this period: the *Andes*, a 212-ton brig, built by Francis Grellier in 1838, owned by C.W. Robin, I.H. Gosset, and John Herault, and destined for the South American trade. This one vessel not only serves as an example of how cod-trade capital was invested in Jersey shipbuilding, but also provides further evidence of why the location of Jersey's shipbuilding industry shifted from the New World to the island in the late 1830s. It suggests that there existed in Jersey a large and growing demand for vessels over and above those needed for the cod fishery, since otherwise a move back

Table 30
New Registries at Jersey, 1834–39

Year Built	New World built			Jersey built		
	Tonnage	Average tonnage	Number of Vessels	Tonnage	Average tonnage	Number of Vessels
1834	436	87.2	5	366	183.0	2
1835	445	148.3	3	259	129.5	2
1836	224	74.7	3	877	125.3	7
1837	384	96.0	4	656	109.3	6
1838	310	51.7	6	617	154.3	4
1839	51	51.0	1	1,054	150.6	7

Source: Jersey Ship Registers.

to the mother country would have been uneconomical. Indeed, the sailing patterns shown earlier not only indicate the importance of the cod trade, but also give a picture of general growth in the carrying trades. It was, in fact, in the nature of the cod trade to create an entry into the carrying trades by virtue of the trading structure involved in marketing and supplying the staple trade. This is reflected in the diversification of the cod-trade firms into other trades, usually by market region and/or in product lines (that is, as a result of deciding what could be purchased in exchange for cod, or to supply the cod-trade settlements using the cash acquired in the sale of the staple).

The seasonality of the cod trade, and how it fits into other trade seasonalities, underlines this point. Figure 25 shows the seasonality of Jersey trades for the sample years 1830 and 1840. It shows the number of vessels that were at ports of call in various trading areas on a month-by-month basis, for the principal trades of the island. In both years, the Jersey trades fit into the cod trade in terms of seasonality. Cod-trade "highs," for example, occurred during Mediterranean "lows," and coasting peaked after the cod trade did, as one would expect. The pattern is that of an increasingly integrated system of supply, production, market, and trading opportunities.

The commodities sought in the trades into which the cod firms diversified are also instructive. In Jersey in March of 1830, for example, DeQuetteville was selling Lisbon salt for the cod trade, figs and wine for local Jersey consumption. In August he was selling Rio coffee, white sugar, and Sicilian wine, and in September produce from the fisheries.[35] Given the nature of inputs into the cod trade, the markets in which the fish was sold, and the freight strategies

A. **1830**

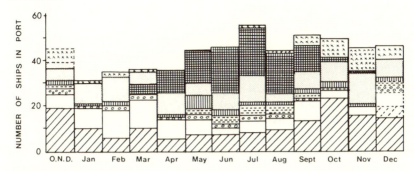

B. **1840**

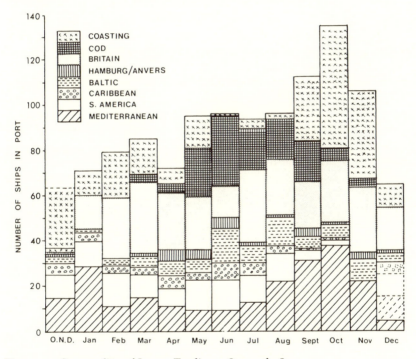

Figure 25. Seasonality of Jersey Trading, 1830 and 1840

employed by Jersey vessels at market, DeQuetteville's commodity
sales indicate an integrated cod-trade system: salt as input, fisheries
produce as output, and figs and wine (Mediterranean) and coffee
and sugar (Caribbean and Brazil) as exchange goods. Likewise, in
August of 1830, LeVisconte was selling Demerara rum,[36] and in

January, Nicolle and Co. were selling Russian hemp for cordage.[37] The Jersey historian Podger commented of this relationship between cod-trade growth and growth in Jersey shipbuilding that "as the cod-fishing trade expanded so new markets had to be found both for the fish and for the produce usually purchased as a return cargo after the fish had been sold. At the commencement of the 19th century it became apparent that wood could be imported cheaply from the Baltic, and with a world in the process of rapid and continuing expansion more and more ships were needed." He also noted the use of Baltic hemp in Jersey shipbuilding.[38]

The seasonality diagrams, then, are a reflection of the logic behind this expanding commerce. They show, along with the old two-way Mediterranean trade, a supply relationship in salt and provisions from the Mediterranean to the cod-trade area, peaking in the Mediterranean in March. This was followed by a production peak in the cod trade, with active shipping from May to August/September. There was also a fish-trade connection from the cod areas into South America related to the Mediterranean and Hamburg/Anvers links, as was shown earlier. It looks as though the South American trades were interrupted by the cod trade in 1830, but that by 1840 that problem had been solved. The market relationship between the Mediterranean and the cod trade corresponds to the September/October Mediterranean "highs," as the cod trade wound down for the year, and the January Mediterranean peak suggests a mid-winter supply trade to Jersey.

These seasonal flows of shipping fitted into one coherent Jersey trading system, as figure 26, representing the voyage diagrams translated into commodity-trade flows, shows. The fisheries sent their produce directly into two of the three groups of countries involved, where it was exchanged for other goods required either in Jersey or in its export markets, or in the fisheries. Fish was sent directly to Honduras, Brazil, and the West Indies, and to England, France, Spain, Portugal, and Italy. From the first three, the fisheries received coffee, sugar, rum, and molasses. From the others, Jersey received a wide range of goods, including manufactured goods, iron, copper, cloth, and so on, from England, and wines, fruits, and salt from the Mediterranean, including France. In Brazil, fish was exchanged either for notes of credit against future purchases to be made by Jersey merchants there, or for coffee, sugar, wine, and brandy sent directly from Brazil to the third group of countries – Russia, Prussia, Denmark, and Germany (Hamburg) – as well as the coffee and sugar for Jersey. From this third group, in exchange for the goods that had been purchased in exchange for fish, came the inputs into the

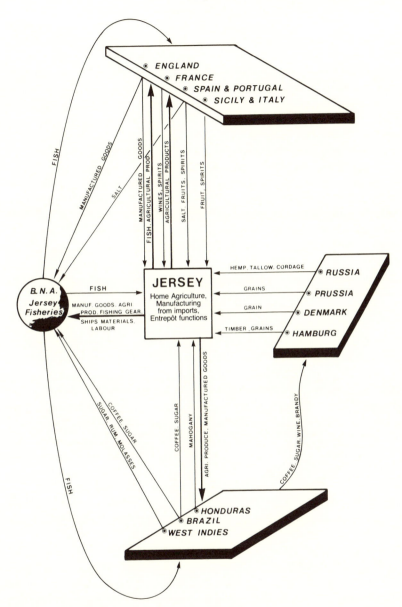

Figure 26. Jersey Commodity Flows, 1830–40

shipbuilding industry: timber, hemp, tallow, cordage, linen. Mahogany came to Jersey from Honduras for vessels, furniture, and house construction. From this last commodity group of countries also came wheat, barley, and "grain" for the use of the fisheries, for consumption by islanders, and as fodder for Jersey cattle. Exports of Jersey produce *per se* were very small – apples, cider, cows, heifers, potatoes, stone, cotton stockings, some flour.

Jersey's major exports were re-exports, purchased with the produce of the fisheries and often further processed in Jersey before sale as inputs into the fishery. Such goods might be purchased directly with fish, or indirectly with such articles as coffee, which had previously been purchased with fish. Hence, Gaspé exported fish to Honduras, for example, and then mahogany was imported from there to Jersey. Jersey's exports, then, were primarily produce of the fisheries, or some commodity derived ultimately from a country which imported from the fisheries. This commodity-flow pattern explains the voyage patterns shown earlier, and it also affirms what the Jersey sales of DeQuetteville, Nicolle, and LeVisconte suggest: that the produce of the fisheries was the foundation of Jersey's commercial wealth. Given the relative insignificance of the island's home produce, its merchants could not have created an Atlantic (and still expanding) entrepôt trade based on cider, apples, cows, and potatoes. Indeed, the immense growth in the coasting trade, from relatively small numbers in 1830 to the largest total number of vessels by 1840, is an important indicator of Jersey domestic development in that decade. It is evidence of local capital accumulation at the level of small (as opposed to merchant) enterprise, the growth of local linkages in domestic business, and spreading effects from the increasing prosperity of the island.

In effect, there appeared to be, in Jersey, generation of the kind of backward linkages into transportation development (albeit still maritime) that was absent in Gaspé. It was the produce of the distant fisheries that gave the island the market commodity, and in sufficient quantity, that it itself lacked. Jersey merchants could then use this produce as a marketable good with which to operate an entrepôt trade (wines, coffee, etc.) and to supply inputs into the island's nascent shipbuilding industry. With the machinery of that entrepôt trade in position, Jersey's expanding fleet could then gain entrance into the carrying trade, and into the wealthy commerce of the British Empire (See figure 25).

The carrying trade, of course, burgeoned with the expanding shipping fleet,[39] and Jersey vessels became the sailing-ship equivalent of the modern tramp steamer, venturing into the world shipping lanes in search of freights. One example is that of the ship *Hasty*, (master, J. LeSueur; managing owner, George Deslandes of St Helier), which in October of 1864 sailed from London to a loading port in Wales, thence to Rosario and then on to Singapore, whence "to all and every safe port ... trading backwards and forwards in any rotation in Asia, Australasia, Africa, America and/or Europe, thence proceed to a port in U.K. for orders to discharge in the U.K. or continent of Europe and finally to proceed to a coalport to load for Jersey" – a voyage of thirty-six months, wages to be paid at the end of the trip.[40] Another example is the ship *William*, which in 1872 left London for Montevideo, whence she was to sail to any of the ports on the east or west coast of South America and the New Zealand colonies, the India and China seas and straits, Japan, the North and South Pacific and Atlantic oceans, the Red Sea, the Persian Gulf, the West Indies, the United States (between Portland and Galveston, inclusive), the British North American provinces, and the Mediterranean, "to and fro as employment may be found."[41] Nor did this exclude the cod trade. The *Canada* (John P. Carrel) of St Helier, a 156-ton vessel, sailed in March of 1880 "to La Poile and thence to any place or places in the North and South Atlantic Oceans, Dominion of Canada, United States of America (between Galveston and Portland both inclusive), West Indies, Brazil, River Plate, the United Kingdom and the continent of Europe (including the Mediterranean) to and fro as freights may offer for any period not exceeding two years and back to the port of Jersey for final discharge." Her crew was to be under the control of DeGruchy, Renouf, Clement and Co. or their agents in Newfoundland.[42]

Jersey ships sought freights of wheat and timber in Dantzig; sugar, rum, and timber in Demerara; beans, grain, cotton, flax, and sugar in Alexandria; cotton in Bombay; tallow in Constantinople; hides and wood in Melbourne; grain in Odessa; they carried coal, fish, wine, bricks, cordage, and countless other articles of trade from Jersey and Britain during this heyday of the British Empire. Commercial capital flourished, and large stores became established, many carrying names long associated with the cod trade. DeGruchy, Le Mesurier, and LeGallais, for example, were also, after 1835, associated with shipbuilding and ship ownership.[43]

The cod fisheries also provided Jersey with a major export market for home produce, as well as for such foreign produce as was needed, for example, cordage, canvas, and so on, from Russia. More importantly, they provided the island with a major stimulus in the form of a large market for home manufactures. In other words, the final-demand linkages of the staple trade (providing consumption goods for people working in the production of the staple) were captured by the metropole. Jersey could by law export to Britain only its produce *and* manufactures – that produce of the island which was manufactured on the island. However, to the fisheries Jersey

Table 31
Jersey Exports to the Fisheries, 1830s

	1833	1834	1835
Potatoes (tons)	732	586	325
Flour (tons)	196	178	312
Biscuit (tons)	257	273	237
Pork (barrels)	760	928	–
British Salt (tons)	447	1,318	395
Foreign Salt (tons)	420	288	722
Bricks (tales)	70,900	21,500	39,450
Cider (gallons)	6,762	2,155	8,400
Sailcloth (yards)	7,531	7,829	8,963
Ready-made sails (yards)	4,493	4,913	6,552
Cottons (shirts, etc.) (yards)	19,653	17,026	16,589
Cloth (articles)	341	53	97
Woollen clothing (articles)	2,978	2,866	2,662
Linen clothing (articles)	3,864	3,743	2,384
Worsted clothing (articles)	2,337	2,005	1,629
Boots (pairs)	1,013	871	705
Shoes (pairs)	12,271	11,309	10,598

Source: *The Guernsey and Jersey Magazine*, 310.

could export the produce *or* manufactures (i.e., regardless of where
the raw materials were acquired) of the island. The distinction was
important:

For while we cannot, in the former case [i.e., England] manufacture a commodity from foreign articles for a free importation into England, we can in
the latter [i.e., the fisheries], which enables us to support our establishments
with more facility, by supplying them with flour and biscuit made here
from foreign wheat and with other articles, at a cheaper rate than we otherwise could: but it must also be mentioned, that all articles for the use of
the fisheries can be imported there duty free.[44]

Tables 31 and 32 show the amount of this export to the fisheries
in 1833, 1834, and 1835, the time of McConnell's *Report* on Gaspé,
and some estimated values of these exports. What is immediately
obvious is that even potatoes – the "prime crop" of Gaspé – were
imported from Jersey, where potatoes were also a local agricultural
product. All other commodities, except salt, had been processed to
a degree in Jersey before export, and even the flour barrels were
made in Jersey, as were the tubs required for the packing of codfish
for the Brazil market.[45] Cloth was usually a finished good, rather
than material for home production in Gaspé: cotton, "cloth," and

Table 32
Approximate Values of Selected Exports, 1835

Export	Approximate Value (£ Sterling)
Potatoes	490
Flour	3,000
Biscuits	2,000
Cider	80
Boots	630
Shoes	2,120

Source: *The Guernsey and Jersey Magazine*, 310.

so on, comprising shirts, hose, handkerchiefs, shawls, counter-panes, trousers, jackets, frocks, mittens, gloves, and so on. As an observer remarked, "The preparation ... of the wearing apparel, gives employment to several persons, and to many during the winter evenings, in addition to their usual daily occupations. Most of these goods are sold to the fishermen resident at British North America, and go toward paying their wages, or the price of the fish which they catch."[46]

Consequently, either directly or indirectly, the cod fisheries created the demand for vessels and for goods (both producer and consumer) that in turn created the expansion of Jersey mercantile interests out of the simple cod-trade triangle into an extended overseas pattern of commerce and trade. Jerseymen were therefore justified in remarking that "our fisheries are not only beneficial from the capital and industry which they directly employ, but they are the means of increasing and supporting other valuable branches of our commerce and industry"[47] ... particularly when the linkages of shipbuilding (table 33) and its ancillary trades are considered in conjunction with the home manufacturing industries that arose out of the processing of inputs into the Gaspé production area (see figure 27).

In the last chapter, it was shown that no final-demand structures in Gaspé resulted from the cod fishery so long as the firms maintained the truck system and their import/export monopsony. Instead, savings generated in the business either were re-invested in the staple-production unit, or else they accrued to Jersey as wages, personal wealth, or capital accumulation. In 1837 it was noted that "most of the men who go over from Jersey in the beginning of the year, to be employed either as landsmen or fishermen ... return home at the latter end of the year ... While they are absent, their wives and families are not idle ... they knit lamb-skin stockings,

Table 33
Trade Relating to Shipbuilding, St Helier, 1843–1890

Year	Shipping Agents	Navigation Teachers	Block Makers	Boat Builders	Anchor Smiths and Iron Founders	Ships' Brokers	Coopers	Mast and Oar Makers	Sail Makers	Rope & Twine Makers	Ship-Builders	Timber Merchants	Merchants
1843	–	–	7	–	5	–	–	–	7	7	7	–	52
1845	–	3	7	–	–	6	–	–	5	7	10	9	63
1850	–	–	–	–	–	–	–	–	–	–	–	–	–
1855	11	9	–	2	–	6	–	–	5	–	12	7	47
1860	11	2	5	–	–	as agents	4	5	5	7	–	–	39
1865	11	–	5	10	7	–	5	5	5	6	11	8	–
1870	10	5	4	6	12	7	6	4	5	4	11	8	31
1875	7	5	3	5	9	6	6	2	3	5	11	8	30
1880	7	5	3	6	10	5	3	3	2	5	9	7	26
1885	9	5	4	6	7	–	4	2	3	5	7	7	35
1890	9	4	–	6	8	8	3	2	3	–	5	5	29

Source: Proudfoot, from the British Press and Jersey Times Almanachs, 1843–90.

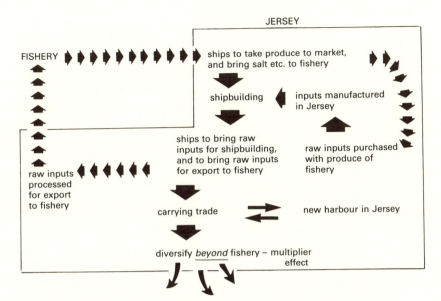

Figure 27. Backward Linkages from Fishery – Jersey, 1830–60

mittens and jackets for them, a great portion of which these fishermen sell previously to their return home, at a profit."[48] Thus, the
need for basic clothing in Gaspé also produced a *cottage-industry*
response in Jersey, the profits from which returned home to Jersey
with the fishermen over and above their wages. Crew lists for
Jersey[49] often showed wages of seamen as being paid on return to
Jersey, thereby enhancing final-demand structures at home rather
than at the fishery, not a surprising situation when it is considered
that many of these fishermen were seeking supplementary income
for a wife and family in Jersey.

Conspicuous consumption, in classic fashion, took the form of
large "cod houses," as they were called locally – the island equivalent
of the West Country mansion.[50] While not built in the "grand manner" of its Dorset counterpart,[51] the Jersey cod house was a spacious
building, often a farm, equipped with coach house, orangerie,
lawns, and functional farm buildings. "Petit Menage," the home of
Raulin Robin, was a moderate-sized farm, valued in 1886 at about
£10,120, including contents. Insurance on it in 1868 had amounted
to £3,000, and had included the house and adjoining offices, the
household goods – books, clothes, and so on – outhouses, coach
house, farm buildings (potato and cattle stores mostly), and cottages
occupied by labourers "in service of the assured."[52] But the Robin

family was not given to ostentation. James Robin's funeral, for example, was described as a quiet affair, and the man himself as "a man who lived out of local politics beyond being a steady friend of reform."[53] Perhaps the best assessment that can be made when considering conspicuous consumption in Jersey is that the Jersey merchant was more enamoured of capital accumulation than of extravagant display.

Capital accumulation in Jersey from the cod trade was not inconsiderable, as the earlier review of the Jersey economy has suggested. One estimate of its value to Jersey in the 1830s amounted to at least £100,000 per annum, and the value of cod imported into Jersey alone in 1837 was over £40,000.[54] The personal capital assets of Raulin Robin at the time of his bankruptcy, in 1886, amounted to £41,960. With personal debts removed, his estate was valued at £31,233 prior to bankruptcy proceedings, including £27,925 which he had advanced to PRC and CRC.[55] In 1833 McConnell had remarked that "the Messrs. Robin have deservedly acquired family fortunes in this district," and of P. and F. Janvrin he observed that "the extent of their business and amounts of capital are unlimited."[56]

The picture presented in an examination of capital accumulation is one of merchant capital flowing to the core from the periphery. The Hon. R.B. Sullivan commented that "the profits of the great branches of Canadian trade ... have become fixed capital in the Mother Country. The merchants who have accumulated fortunes here ... have generally returned home to enjoy the fruits of their labour."[57] In Jersey, however, the capital was often re-invested, and sometimes to a considerable degree, if Raulin Robin's advances to CRC and PRC of £27,925, that is, 89.4% of his estate in 1886, is typical of the investment pattern among Jersey merchants of the time.

However, the pattern of Jersey re-investment of capital accumulated in the cod trade warrants further research, as indeed does the whole question of linkage effects from staple trades in the "core." Detailed investigation might uncover the nature of re-investment and help to identify shifts in investment from merchant (staple) sectors to industrial or other sectors of the core economy, thereby allowing a finer appreciation of the role of colonial staple trades in the mother-country economy than can be accomplished here. Certainly, very little in the way of genuine industrial capital investment occurred in Jersey, apparently not so much because of reluctance on the part of investors (the oft-cited conservatism of merchant capitalists), but because, in terms of its size and location – forty-five square miles and close to Britain – Jersey could not compete with its giant neighbour in external or internal markets for industrial goods.[58]

It is not surprising, therefore, that along with increased capital accumulation and its re-investment in small manufactures, wooden shipbuilding, and commerce came the creation of finance capital in Jersey. The firm of Janvrin, Durell and Co., for example, was the "Commercial Bank," and grew directly out of the cod trade. In 1816, the firm of P. and F. Janvrin began selling off its sailing vessels and stores, and in 1817 it was listed in the Almanachs of Jersey under "Banquiers de la Ville."[59] In 1841, Frederick Janvrin sold the firm's Gaspé establishments – houses, stores, wharfs, cookrooms, and so on, at Grand Grève, Gaspé Basin, Malbaie, Cap Rosier and Griffon's Cove – in one lot, with inventories and debts.[60] They were sold, of course, in Jersey, as was the custom,[61] and with the sale the Janvrins completed their move from merchant to finance capital. This bank became Robin Frères in later years. Other banks resulted from similar shifts: the Old Bank (Hugh Godfray and Sons), the Mercantile Union Bank (Nicolle, De Ste Croix, Bertram & Co.), and the Jersey Joint Stock Bank (Mathews, deCarteret & Co.).[62]

This makes sense. Wealth in Jersey in this era came from the sea and the shipping lanes; apart from investment in the one "industrial" sector – wooden shipbuilding – capital tended to be re-invested either in the staple trade or in the commercial and, later, financial sectors of the economy. The "trade and navigation of the Island," which the Chamber of Commerce fought so determinedly to protect and encourage, could never have been achieved on the basis of local produce: the far-flung world-trade linkages seen in such firms as Nicolle et Cie were inherent in the cod trade. Indeed, the effective ocean "harvest" from the fisheries and the related shipbuilding and carrying trades – the legacy of the cod merchants of Jersey – implied a minimal interest in land-based development, be it in Jersey or at the fisheries. It did, however, encourage more ephemeral skills; the island's evolution toward an economy which was increasingly dependent on "invisibles" (the returns from finance capital and shipping) seems both sensible and natural when seen in this context.

In summary, then, when recalling the trading geometries of Jersey in conjunction with the non-expansion of the Gaspé economy, the expansion of the island economy demonstrates the purpose of colonial states and metropolitan monopoly structures. Put bluntly, from the island's perspective, the value of the Gulf colony was in the development of commerce and wealth in Jersey. However, the necessity of limiting or containing expanding linkages in the Gulf region, while advantageous to the Jersey economy, was first and foremost the cod-trade merchant's strategy for protecting his supply functions on the coast (so as to retain his "productivity" in terms of quintals of fish sold at market, and hence the profitability of his firm) rather than the result of any conscious altruistic or patriotic desire to expand the Jersey economy at large.

Finally, it is now possible to identify more precisely the dual effects of the cod trade on colonial Gaspé and on Jersey. The weakness of forward linkages can be clearly attributed to the staple itself, since further processing appears neither in Gaspé, nor in Jersey, nor in the markets for the staple. The backward-linkage effects, however, are a different matter. The lack of transportation development on land, in the form of "no roads," can be attributed to the nature of the staple; but the maritime equivalent, in the form of the building of vessels to sail the sea-lanes, should have resulted in Gaspé from the staple industry – and indeed for a period of time it did, providing a large part of Jersey shipping in the early years of the nineteenth century. Consequently, the backward linkages of the shipbuilding industry should have also accrued to Gaspé, at least to the extent that the industry serviced the cod trade. But blacksmiths, carpenters, and the like were transported from Jersey, as was much of the equipment that was input into the industry – sails and anchors, for example. Boatbuilders (the Mabé family) were also supplied from Jersey,[63] and so no real backward linkage developed through their enterprise either. The extent of the increasing loss, especially after 1840, can be estimated from the extent of the growth of ancillary trades in Jersey, always keeping in mind that a parallel development of the carrying trade cannot be assumed with any degree of certainty, and therefore a shipbuilding "boom" of the kind experienced in Jersey cannot be claimed. Thus, the harbour improvements in Jersey, which were (at least in part) a result of the growth of shipbuilding, were mostly due to the expanding carrying trade, many of the fixed assets of the shipbuilding industry having consisted of wooden slipways and sheds constructed on the beaches of St Helier's and Gorey bays. Much of the employment in the industry was of the "journeyman" type, with carpenters, blacksmiths, and others moving from slipway to slipway as the demand for their services arose.[64] The industry in Jersey, relative to the British steamship industry, was very lightly capitalized, and fixed property was minimal. It cannot be argued that to claim such a loss for Gaspé, however, is totally unrealistic, since a wooden-sailing-ship industry of considerable proportions did arise elsewhere in Atlantic Canada during these years. From Quebec City round the Gulf and north to St John's,

Newfoundland, as well as south to Saint John, New Brunswick, the construction of such vessels became an important sector of several local economies, and the total fleet of the region became the fourth-largest carrying-trade fleet of its kind in the world by 1867. The structure of the Canadian industry was very similar to that found in Jersey.[65]

Final-demand linkages should also have accrued to Gaspé from the production of producer and consumer goods for fishermen on the coast. Some idea of the cash value of these goods can be gleaned from the estimated values given earlier for some of the goods imported into Gaspé from Jersey. The manufacture of those goods shown in table 31, however (comprising mainly foodstuffs, clothing, and fishing tackle), was confined to Jersey, and the home industry flourished accordingly. As time went on, though, industrialization in England resulted in the importation of British manufactured goods from Liverpool to the fishery, which benefitted neither Gaspé nor Jersey in terms of linkage effects. No significant final-demand structure emerged in Gaspé, however, during the years of merchant import/export monopsony and "truck" control, and capital accumulation and investment were likewise confined to Jersey.

With respect to the benefits arising out of the cod trade which accrued directly and, to a degree, indirectly to Jersey, the merchant system ensured that they did not accrue to Gaspé in order to protect and promote the metropolitan merchant's control of the trade. It is possible, therefore, to speculate about what might have happened had the merchant triangle not operated as it did. At the very least, local demand structures should have materialized in Gaspé, and cottage industry should have flourished, giving some kind of cash basis to the economy. Local agriculture would perhaps also have been stimulated, and certainly local backward linkages from ship-building, to some unidentifiable degree. Local demand structures, combined with increased agricultural production, might arguably have resulted in the construction of roads and the opening up of the hinterland, and this in turn might have created sufficient diversification around the initial staple base to have generated some expansion of the domestic Gaspé economy. Whether Jersey could have developed the entrepôt function which enhanced its carrying trade without the linkages from the cod trade is an unanswerable question ... but it would certainly have been a more restricted trade, if it had occurred at all. There is no doubt, however, that the claim that Jersey's cod trade functioned as the springboard for the island's development in the nineteenth century is justified.

CHAPTER SEVEN

The Collapse of the Merchant Triangle

There occurred at mid-century a series of events that would in time combine to upset the balance of the merchant triangle. The underlying causes of most of these events can be found in the progress in Britain and western Europe of the industrial revolution, which changed British economic structures and political strategies. The marginality of Jersey and, even more so, of Gaspé left them powerless to redress the imbalances that resulted from the shifting political and economic philosophy of Britain. In the past, the Jersey merchants had been able to withstand stress at one, or even two, of the apexes of their triangular system of trade; the mid- to late nineteenth century was to witness disturbances first at one, then at another, and finally at all three points of the merchant triangle. This the system could not withstand.

The first upset occurred in the market apex of the triangle with the British West Indies' loss of control of the British sugar market in 1852, when the protectionist duties on sugar were repealed.[1] This was not major in and of itself. The West Indies market had proved insecure in the past, and the emancipation of slaves had disrupted that trade, as was noted in chapter 6. Hard on the heels of this came the signing of the Reciprocity Treaty with the United States in 1854, which, as noted in chapter 5, was demonstrably unfavourable to the inshore merchant fishery, which had the chronic problem of trying to protect a resource base with an unprotected rent. In 1886, Mitchell, the fisheries minister, commented of the Treaty that "les privilèges de pêche cédés aux Américains, et l'omission de privilèges semblables aux Canadiens ... les termes du traité de 1854 en ce qui concerne les pêcheries sont assez favorables aux États-Unis."[2] The cod merchants had been no happier. Indeed, at the time, the Chamber of Commerce in Jersey had begged Parliament not to alter the existing treaties, dragging out all of its tried-and-tested arguments on tariffs and bounties that militated against the British North America fisheries in the markets, and reiterating the old "nursery for seamen" dictum that had been so successful in the earlier era of a mercantilist British government.[3] The petitions failed, and the cod merchants of Jersey were left to cope with American competition as best they could. This they did by tightening their credit system until the Treaty was abrogated in 1866 (see chapter 5).

Development in Gaspé was still retarded at mid-century. Despite considerable population increase, some growth in agricultural production, and the appearance of a small degree of processing of local foodstuffs and timber (see table 34), 85.8% of all occupied land remained under wood or in its "wild" state, and 58% of all houses were still of log construction. There was only one gristmill per 211 households and one sawmill per 169 households; potatoes remained the primary crop of the area. In the late 1850s, however, the population was swelled by a steady flow of migrants not attached to the merchant houses, who began settling the area to farm in such places as the Matapedia Valley. At about the same time, population pressure and a deteriorating economy and environment in the upper parishes along the St Lawrence River (particularly in the counties of Kamouraska, L'Islet, and Montmagny) induced many French-Canadians to settle permanently along the northern shore of the Gaspé, from Cap Chat to Gaspé Basin.[4] By 1860, the population of the Gaspé District had risen to 27,169 persons, and that of Cap Chat to Île de St Barnabé (Rimouski) to 10,000 – a total growth rate of 270% over the period from 1830.[5] Indeed, ten years earlier, CRC had noted of Grand River that the population there was now too large to be supported solely by the fishery.[6]

In 1860, Gaspé Bay was made a free port, and a local market was thereby instituted. The *Boston Commercial Bulletin* reported:

A free port at Gaspé Basin ... This Act will afford the same advantage to all nations. As a consequence of this free trade measure, the inward navigation *via* the St. Lawrence and the Lakes is also declared free of all charges, so that our vessels may find their way from this port to Chicago and back, with either the fish of the sea or the manufactured goods of Massachusetts, or both, and return with flour or pork or other commodities. Then, again, our fishing vessels need not return with each 'take', but can carry them into this free port and sell them to other vessels from Europe or South America, coming to that place for them in return for the products of their respective nations, or for cash ... Our fishing firms could make Gaspé their headquarters in the Bay, and could send an agent there to furnish supplies ...

Table 34
Gaspé County, 1851–52, Economic Development

Social Information

Houses Inhabited	Number of Families	Houses built	Shops and stores	Schools	Churches	Acres of land occupied	Persons Owning Land
1 brick 628 frame 870 log	1,688	86	104	10	26	92,210 4,944 – crops 7,875 – pasture 322 – gardens 79,069 – wild & wood.	1,497

Agricultural Produce

	Wheat	Barley	Rye	Peas	Oats	Potatoes	Turnips	Clover/Timothy
Acres	461	1,116	273	210	890	1,016	240	–
Bushels	3,418	10,850	1,242	1,148	9,962	67,114	10,751	15

Animals and Animal Products

Bulls/Oxen Steers	Milch cows	Calves/heifers	Horses	Sheep	Pigs	Butter	Cheese	Beef	Pork
1,122	1,937	1,090	826	5,543	2,597	56,604	1,271	610	740

Manufacturing

Flax/Hemp (lbs)	Wool (lbs)	Maple sugar	Fulled cloth (yds)	Fulled linen (yds)	Fulled flannel (yds)	Grist mills (water)	Mill returns	Saw mills (water)	Mill returns
838	12,020	14,310	1,288	503	6,113	14; 8 men employed, 6 mills, no returns	4,765 barrs. from 8; 9 returning capital invest.	19; 28 men employed, 9 mills, no returns	8 – 146,000 feet; 2 – 15,000 planks; 17 – capital invested, returns of £803.

Source: Census of Lower Canada, 1851.

thus saving the time spent by the fishermen in returning home from their first trip.[7]

In 1862, Fortin reported 258 Acadian immigrants working fifty-four farms in the Matapedia Valley, and added that more arrivals were expected the following spring.[8] The northern coast of the Gaspé was likewise experiencing a shift from an economy totally dominated by the fisheries:

The population of our county is chiefly composed of fishermen ... and, in truth, in no part of the province has agriculture been more neglected, and nowhere is it so little appreciated, as in our county, and the eastern section particularly. However, there is perceptible of late years, amongst our young habitants, an anxiety to acquire the ownership of the soil, and to develop its resources; a tendency which the bad success of the fisheries for some time back and the increase in the population, will further stimulate.[9]

Although many of the French-Canadians who settled this region of the peninsula, particularly in the 1860s, participated in the fisheries, there was a distinct influx of immigrants who were solely preoccupied with farming. A new interest on the part of French-Canadian politicians and clergy also threatened the Jersey merchants. Government policy – especially after 1867, when the newly established Provincial Government of Quebec created a special "Ministry of Colonisation" to promote the expansion of agricultural settlement throughout the province, including the whole of the Gaspé – was not sympathetic to either the fisheries or the fish merchants.[10]

The opportunity to trade with persons other than the local merchants and the establishment of a genuine agricultural base were then enhanced by the construction of the first road in the area: one of the "colonisation roads," which were government funded. Taken together, these developments meant that the economy at last had a basis for diversification, and that the era of Jersey hegemony on the coast was coming to a close. Fortin, aware of the implications, exulted that "c'était une ère nouvelle pour ce pays: aussi ne tarda-t-il pas à faire des progrès rapides ..."[11] The situation did not, of course, destroy the merchant fishery. Gaspé access to the staple markets was still provided by the Jersey sailing fleets, and it was still the Jersey merchants who alone possessed the skill in market manipulation that came with long experience, personal contacts, and the vertical integration of their firms. These factors ensured that the old fishing firms retained a considerable degree of mastery over the Gaspé fishing economy. But mercantile control of the coast was weakening, and firms such as Janvrin left the coast and the business in these years.[12] By 1865, Fortin could rejoice that "Ce n'est plus maintenant (et ce ne peut plus l'être) la manière de commercer d'autrefois ... Or, il est à ma connaissance qu'il est fait cette année bien de ventes de poisson pour des milliers de louis, argent comptant."[13]

After 1860, the cod-trade system also began to experience increasing strain in its lines of communication between the three control points. The rapidity and regularity of steamships was beginning to seriously threaten the hegemony of sail in the carrying trades.[14] In 1870, a *Mitchell's* editorial commented that in 1869 the total tonnage of new steam vessels registered in Britain had been equal to nearly half the tonnage in sail. They observed that whereas in 1865 only 0.22% total tonnage of new vessels in that year had been constructed of iron, four years later the new wooden vessels registered had dropped to 38.8% of the total tonnage of 1863. The paper concluded that "these returns disclose the revolution that is taking place in shipping, for iron-built vessels are slowly but surely superseding wood, and steam is supplanting sail ... We need not say that this employment of steamships must have tended to retard the increase of sailing tonnage and keep down freights."[15] In Jersey this was clearly the case, and the island's sailing-ship carrying trade, a vital adjunct to the cod trade, was collapsing under the new technology. In March of 1870, *Mitchell's* printed a report from Bombay which warned that "what had taken place in the Mediterranean is now being repeated in India. The Rivalry between steam and sailing vessels has fairly commenced, and should the Suez Canal remain open for navigation of which there can now be no doubt, it seems probable that in a few years the whole trade of this port will be carried on by steamers."[16] It was becoming clear that any apparent reversals of the trend were no more than "the effect of a temporary and very limited demand for some particular ship on the point of sailing."[17] In June of 1870, reports from *Mitchell's* warned that even the Buenos Aires freights were being taken over: "Sailing vessels find it difficult to obtain freights there now there are 4 1st class mail steamers plying to that river."[18]

Many traditional Jersey markets were being lost to the Jersey sailing-ship fleet. Table 35 gives some sample figures for the fall in freight rates that accompanied the fall in prices between 1860 and 1870. It also indicates the losing battle of sail against steam in the carrying trades. The 1867 Annual General Meeting of the Jersey Chamber of Commerce, for example, referred to the poor situation of trade generally, and to the depression and collapse of many shipping interests. In Jersey, the effects were being felt as, at the very

Table 35
Freight Rates, 1860 and 1870

From/To; Commodity	1860	1870
IN to Britain (London):	from West Indies:	West Indies:
Sugar	47/6 to 65/–	21/6 to 40/–;
		60/– from Manila
mahogany etc.	55/– to 60/–	37/6 to 40/–
OUT from Britain (London):		
coal	40/– to Buenos Aires	21/– to 30/– to Brazil
		and West Indies
machinery	60/– to Buenos Aires	30/– to Brazil
		and West Indies

Vessel Type

STEAM freights		
to Britain (London)	45/– to 50/– to Patras	45/– from Buenos Aires;
SAIL freights		
to Britain (London)	35/– to Patras	60/– from Manila;
		34/6 from the Danube

Source: *Mitchell's* 1860, 1870.

least, a reduction in profits, despite the fact that total tonnage on registry in the island was now 51,000 tons. The Chamber referred to "the unremunerating character of the Newfoundland trade," and observed that shipbuilding in Jersey had virtually ceased in recent years.[19]

The decline of Jersey shipping is clearly seen in the ship registers. Table 36 shows newly built tonnage registered at Jersey by quinquennium from 1850, except for the year 1855, for which data are incomplete. Newly built tonnage is a measure of new investment in the shipping sector, and it is clear that this peaked in the mid-sixties and then fell off dramatically, whether considered by total tonnage, mean tons, or number of vessels. By 1880 there was no merchant investment in new vessels (table 37) and the range of vessel tonnage held by owners had shrunk steadily, and there was no new shipping investment whatsoever by 1885. The golden age of Jersey seafaring was at an end, and its cod fishery, so intimately associated with that maritime economy, was deeply troubled.

By 1873, prices in general had begun to fall, and the drop continued until the end of the century. This Great Depression was in fact a series of booms and slumps, on which there is an extensive literature.[20] Sugar, petroleum, cotton, tea, silk, wheat, iron, and steel

Table 36
Newly Built Tonnage at Jersey[1]

	1850	1855[2]	1860	1865	1870	1875	1880	1885
Total new tons	1,300	–	1,786	4,442	783	901	86	–
Mean new tons	81.25	–	63.78	130.6	87	81.9	43	–
No. vessels	16	–	28[3]	34	9	11	2	–

Source: Jersey Ship Registers
Notes: [1] 1855 data incomplete
       [2] All tonnage is gross tons
       [3] Including one paddle steamer

particularly suffered from falling prices; many of these commodities were vital parts of the Jersey carrying trade. Sugar in particular remained in serious trouble, not only because of the removal of protectionist duties (which had occurred in 1852), but because the West Indies was a victim of poor production in the plantations, British laissez-faire policies, and the industrial production of beet sugar. *Mitchell's* had remarked in 1860 that "the increase in the growth of beetroot sugar on the Continent has been a most rapid one ... we regard such a [industrial] system of production as injurious to Commerce, and highly prejudicial to the interests of our Colonial dependencies ... Foreign sugars ... have displaced West India produce to a serious extent."[21] Williams has commented that beet factories in Europe developed rapid production – "11 factories in Germany were needed to produce what Barbados' 440 exported"[22] – but he noted that the battle was being fought primarily in the British marketplace, and he blamed the 1852 repeal on duties:

It was the highwater mark of Britain's free trade policy. Thereafter Britain's interest was to buy in the cheapest market ... Thus it was neither science, not technology, nor availability of capital, nor the state of profit which explained the triumph of beet over cane. It was Britain's free trade policy and her desire for cheap sugar. It was that policy and that desire which gave rise to and stipulated the greatest mockery and the very antithesis of free trade, the bounty system ... The British public, refiners, manufactures of jams and candies benefitted.[23]

In Jersey, the collapse of the sugar market, coupled with the new ship technology and the resultant decline in freights, created a total dislocation of the local economy. The first major portent of the trou-

Table 37
Vessel Owners and Vessel Size, Jersey, 1850–80

	1850	1855[1]	1860	1865	1870	1875	1880
Shipbuilders	4	–	5	4	1	1	–
Mariners	9	–	13	12	5	5	–
Merchants	1	–	7	11	2	3	–
Gentlemen	2	–	2	–	–	–	–
Ironfounders	–	–	1[2]	–	–	–	–
Tradesmen	–	–	–	3	–	–	–
Shipowners	–	–	–	2	1	1	1
Landowners	–	–	–	–	–	–	–
Other	–	–	–	2	–	1[3]	–
Range of tonnage owned	15 to 432 tons	–	4 to 321 tons	21 to 619 tons	24 to 275 tons	45 to 202 tons	7 to 79 tons

Source: Jersey Ship Registers
Notes: [1] 1855 data incomplete
        [2] This vessel was a paddlesteamer of 65 tons with one mast
        [3] Potato merchant

bled times ahead was a financial crisis that occurred in 1873, with trade in a state of "partial stagnation." The Jersey Mercantile Union Bank collapsed, and shortly thereafter so did the Joint Stock Bank. DeQuetteville of Labrador failed in that year, directly as a result of the crash, which was itself a result of overexpenditure on new harbour facilities at St Helier, but indirectly because of outmoded business methods and the impact of legislation (see chapter 3) which penetrated the import/export monopoly of the firm on the Labrador coast.[24] Similar financial crisis in England had resulted, in 1858, in the passing of an act providing for the adoption of limited liability, but it was rarely implemented until 1873, when the failure of the City of Glasgow Bank (also an unlimited joint-stock bank) forced the introduction of reserve liability.[25] The response in Jersey, in the teeth of emigration and distress, was for the Chamber to recommend the suppression of private bank notes and the publication of lists of shareholders and issues of notes. They also suggested that the States should control money, dispensing it to banks in return for adequate guarantees.[26]

But worse was yet to come, and the depression showed no signs of slackening. The situation was reviewed in the 1879 Annual General Meeting of the Chamber of Commerce. The members agreed that Jersey was suffering from the "universal depression of trade," as well as from a "continued deprivation of the shipping interests"

whose prospects for improvement were, at best, remote. They complained that they had no new source of industry with which to fill
the gap.[27] The following year, they explained that Jersey could not
expand much, because "apart from our shipping interest and fisheries we are an agricultural and domestic community."[28] They therefore looked, as in the past, to the fisheries, which had "ever been
a most important factor, the prosperity or failure of which is shared
or lamented by the whole trading community."[29] The only bright
spots – if they could be called such – were a slightly better fishing
season than had been the case in previous years, and the fact that
the island was small. This, they observed, militated against the development of "any great commercial enterprise or industry," but at
least allowed them to "avoid the universal depression."[30]

They adopted a resolution to develop agriculture instead. The
following year, with the policy already put into effect, the tourist
industry also began to expand, and looked as if it would become
intimately connected with the island's prosperity.[31] By 1884, island
potato production for export had increased, and the number of tour-
ists was also rising; Jersey had exported £282,000 value in head of
cattle to the United States and Britain (47,000 tons) and had commenced the export of fruit and flowers.[32] In 1885, the Chamber
reported a return of £343,756 on potatoes, and was pressing for
better-quality produce.[33]

At the beginning of 1886, the final blow fell for the old Jersey fish
merchants. In January of that year a bank crash destroyed the financial basis of the two greatest cod-trade firms in Gaspé, CRC and
LeBoutillier, as well as of other smaller concerns. On the 11th of
January, 1886, the *Nouvelle Chronique de Jersey* reported that the Jersey
Banking Company had suspended payment. The direct cause of the
crash was mismanagement of funds by Philip Gosset, who was both
manager of the bank and treasurer of the States of Jersey. Public
funds were thus also involved: the bank was known throughout the
island as the States Bank. The financial well-being of several large
commercial houses on the island was in jeopardy, since large sums
of money had been advanced to them, and the crisis was all the
more severe since it was an unlimited-credit bank. The list of shareholders published on January 11th included nearly all of the established families involved in the cod trade and associated concerns:
names such as DeHeaume, Gosset, Payn, Clement, Nicolle, LeBrocq,
and De Ste Croix.[34] As a result of the linked nature of the cod-trade
firms and their associated banks, Robin, Gosset & Co., Nicolle &
DeHeaume (a Jersey-based company), LeBoutillier Frères, whom the
latter represented, and the bank owned by DeGruchy were all threat-

ened. The Commercial Bank (Robin Frères, owned by Philip Snowden, and C.J. Robin) was not involved since it was not part of CRC,
and this branch of the Robin family later purchased the land property
of Raulin Robin when the latter was declared "en désastre" after the
crash.[35]

The fall of CRC was bewailed throughout Jersey as a major tragedy.
The principal forums of informed opinion – the newspapers and the
Chamber of Commerce, of which Raulin Robin had been President
– noted the vital part played by the cod trade, and CRC in particular,
in the economic well-being of the island for more than one hundred
years. In Paspébiac, riots erupted when the news broke,[36] and the
stores were looted by local planters and fishermen, for whom the
firm had long been their only channel of export and hence their
means of livelihood and sustenance.

After 1886, the Minute Books of the Chamber never again mention
the cod fisheries. By 1887, Jerseymen were commenting that, bad
as it had been, the island had survived the crisis[37] and could look
forward with returning confidence to the years ahead. Increasing
stress was laid on the agricultural and tourist potential of the island,
servicing an expanding London market.[38] Robin Frères continued
in business unharmed until the turn of the century, when it was
taken over by an English bank. After a few months of financial
instability, DeGruchy's department store announced that it could
once again open its doors to the public, and thanked them for their
forebearance in the proceeding months.[39] But CRC and PRC were
gone, their assets liquidated to repay the shareholders of the fallen
bank.

The pragmatic attitude taken by financial and commercial concerns
is best explained by the near-eulogy printed in the *Nouvelle Chronique*
of the 11th of January, 1886. It reported:

Dans un cataclysme financier pareil à celui qu'on déplore aujourd'hui ... il
ne nous appartient ... que de counceiller ... la prudence et la patience ...
On a de respecte pour un général qui perde une bataille. On déplore sa
chute, on laisse tomber une larme de sympathie, lorsqu'on sait qu'il était
engagé dans une bonne cause. Par analogie, donc, ne doit-on pas entourer
de tout le respecte possible, de la plus profonde sympathie, le nom de
'Robin'.

But the newspaper saw the end of an era represented in the fall of
this commercial house, and, perhaps unintentionally, pointed to the
real causes of the crash: "Nos générations a vu des grandes choses.
Les navires en bois remplacés par ceux de fer. La voile remplacé par

la vapeur. La correspondance remplacé par le télégraphe. Les vieilles choses se passent. Toutes choses redeviennent nouvelles."[40]

The merchant triangle had finally collapsed. In Gaspé, a series of bad fisheries had struck at a weakened mercantile base. In the cod-trade markets in Europe, competition from other countries had been steadily increasing, the old sugar and timber markets had likewise been failing, and the old carrying trade had been rapidly dying under the onslaught of competition from steam and iron ships. In the past, the merchant fishery had withstood stress at two of its three control apexes, but simultaneous attack at all three bases proved fatal. The financial-support system in Jersey was crucial to the supply of capital and goods to the triangle's production apex, and with that support removed the system could not survive. Indeed, not only had the three apexes of the cod-merchant triangle come under intolerable stress, but so had the lines of communication between them. What had really happened was that the whole context of the trade had changed, and changed permanently. In the crisis of the 1790s, when war had had a similar effect on the apexes and communication system of the triangle (chapter 2), the disruption and cessation of trade thus brought about was temporary. With peace re-established, the trade could, and did, recommence and flourish. But in the 1880s, the wider economy within which the cod trade operated had been altered forever, and there was no place for such a merchant-capital organization within the new industrial world order.

The firm of CRC, under new ownership and with a simplified business structure, survived, but the Jersey emporium in the Gulf of St. Lawrence was gone. As the old metropolitan core across the Atlantic collapsed in the later years of the nineteenth century, a new metropole developed – not in the Atlantic region, as Halifax businessmen had so dearly hoped,[41] but in Montreal and Toronto. As early as 1850, CRC had complained that "poor Gaspé has always been neglected and will be,"[42] and the second half of the nineteenth century bore out the prophecy. The road and rail systems, when they did come, served not to integrate the area so much as to open up a transportation channel down which labour, and the material and financial resources of the area, drained to the metropole: "The completion of the Intercolonial Railway, the deepening of the St Lawrence waterway, and the extension of railroads to the Pacific, together with the disappearance of the sailing vessel, attracted labour from the fishing industry. Nova Scotia turned to the interior in Canada and the United States ... The results of the retreat were evident in the revolution from an economy facing the sea with a large number of ports to an economy dependent on a central port

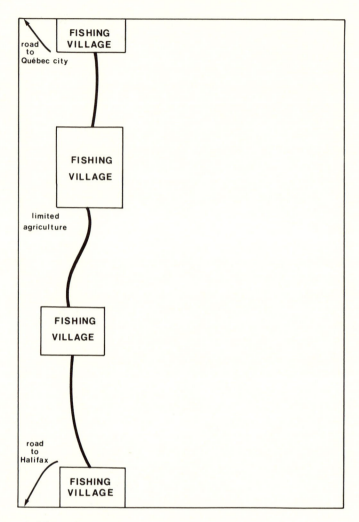

Figure 28. The Fishing Village as a Quebec Outport.

and railways to the interior."[43] The pattern of exploitation remained.
What changed was the direction of the outflow, and the nature of
the goods that were removed. The draining of the profits and produce of the old fisheries, from the coast along the sea lanes to the
old Jersey core, was replaced by the removal of the area's labour
and resources, down through the newer transportation channels to
the new continental core.

With the exogenic support system – so necessary to the export-
based fishing staple economy of the Gaspé coast – removed, the

area turned rapidly in upon itself (see figure 28) to become a rural backwater in the Quebec provincial economy. Arguably, its potential had never been very good, dependent as it had been on the fish staple with its poor growth implications. The capacity to accumulate capital had never been realized, nor had domestic entrepreneurial skills been fostered, given the local control exercised by the Jersey merchants. In this respect, it is significant that when CRC was resurrected, first by Collas and later by Jones & Whitman, the company continued to bring skilled and managerial labour from Jersey until about 1930. At that time, the Depression made the firm uneconomical as a fish business. The necessary marketing skills and contacts required if the local fish economy was to regenerate "on its own account" were lacking, and the "transition from dependence on a maritime economy to dependence on a continental economy," as Innis pointed out, was "slow, painful, and disastrous."[44] It also seems to have been unavoidable. All that remains today of the Jersey merchant emporium is a series of department stores – in Jersey and Gaspé alike.

CHAPTER EIGHT

Conclusion
On Colonial Staple Trades

Once upon a time there were six blind men who set out to examine an elephant. One of them, approaching the animal from the side, declared it to be like a wall. The second, feeling the tusk, pronounced it spearlike. The third, taking it by the trunk, found it to resemble a snake. The fourth, feeling the knee, immediately said that it was like a tree, while the fifth, touching the ear, thought it most like a fan. The sixth, seizing it by the tail, compared it to a rope.[1]

The explanations that we have for colonial staple trades are rather like this. Many studies have grasped part of the truth, but a persuasive overview is rather more difficult to come by. Instead, like the six blind men who promptly fell into dispute over the concept of "elephant," we are heirs to a proliferation of competing approaches to the concept of "staple." This book, in looking at one commodity trade, has uncovered the complementarity of the roles of production functions (land, labour, capital), linkages, colonies, metropoles, and other facets of such trades. It has shown the Jersey-Gaspé fish trade to have been at heart an interdependent system which, under merchant capital, led to the development of Jersey and the underdevelopment of Gaspé. To look at the cod trade is to incorporate everything from the nature of the resource itself, through the fishermen-producers on the Gaspé coast, all the way to the political and economic power of the British Empire ... because that was the context of the trade.

Failure to appreciate this kind of broad context has restricted our understanding of colonial staple development in the past. Partial explanations exist in export-base studies, but further insight has had to be sought in metropolis/hinterland and core-periphery studies, in the work of economic geographers such as Vance, econometric studies such as that of Birnberg and Resnick, and economic histories such as that of Antler.[2] Unfortunately, these schools of thought have often been in competition, rather than seeking cooperation. There has been an insistence now on one, now on another, facet of colonial behaviour as the key explanatory variable, be it class, capital, political or financial power, or economic structure, depending on the perspective from which the problem has been investigated. This has militated against the achievement of a synthesis that would produce a rich and complex analysis more closely reflecting reality. Yet each of these perspectives contains an important part of the truth, and that truth has relevance to our understanding not only of the past, but also of the present. For the roots of the problems that face the east-coast fisheries today lie buried deep in the accretions of two hundred years of attitude formation, perceptions, and behaviour – social, political, and economic.

The Canadian east-coast fisheries are seen today as a "basket case": a drag on the region and on the nation. The head of the MacDonald Commission, for example, stated informally that the fisheries should be abandoned as a hopeless sector, while another member of the same commission expressed the opinion that the only solution to unemployment in the region is emigration. Both of these are depressingly familiar suggestions which underline the demoralization that is dangerously rampant in the eastern Canadian psyche. They echo the frustration of a region which knows that one prime rationale for its settlement was that same fishery, a fishery which in the early centuries proved so lucrative that it drew merchants and immigrants to the area. Yet the industry has been perceived as a "failed staple," despite the fact that it is based on what was until recently a rich resource of cheap protein in a food-hungry world, one which the deep-water fishing fleets of other nations still find economical to exploit. The people of Atlantic Canada are puzzled by this paradox, while the memory of their past glories remains to haunt the region's fishing population with feelings of inadequacy and failure.

This book has presented one case study of that past. To draw together what lessons it has to offer, it is necessary to start at the beginning, with terminology. Buried in the heart of the mythology of Atlantic Canada is the idea that the fish staple had a great deal to do with the development of the regional economy. In the specific case of Gaspé, the Jersey firms (and CRC in particular) carried that burden; but that statement contains an inherent confusion, because a firm is a business, not a staple. Firms are concerned – as this study has shown for CRC – with costs, profit, efficiency, productivity, and the like: that is their legitimate sphere of activity. An aggregate of firms in the same kind of business makes up an industry. In a given unit such as a regional economy, a complex of similar industries makes up a sector of the economy. In other words, there are increasing scales of activity involved which need to be considered separately.

A *staple* is, in the first instance, purely and simply a resource. Resources are, in this most basic sense, neutral: they *are*, and are open to a myriad of different ways of being developed or even of not being developed. They are not yet hooked into any particular socio-economic philosophy. But this is a very simplistic way of looking at staples. Innis, Mackintosh, Lower, and many others[3] endowed the word "staple" with richer meaning. These scholars created a "staples approach," which turned a resource into the leading edge of a new economy at the national, regional, or colonial level, through the development of this resource for export in developed markets elsewhere. This is, in itself, an evolution in thinking out of the old idea of "staple" as the "prime good" of an area, which is the way the term was used in, say, Elizabethan England. It is this Canadian innovation in thinking that led to the famous notion, now a cliché, that the history of Canada is the history of her fish, fur, timber, wheat ... and so on.

And so staple "theory" developed. In Canada, fortunate regions had either a *succession* of staples (British Columbia is the classic Canadian example: fur, then gold, then timber ...), or a *complex* of staples (timber plus wheat in early southern Ontario; timber plus fish plus minerals in early British Columbia). Either of these alternatives allowed well-endowed regions to develop securely because they had an inbuilt "capacity to transform," as Kindleberger puts it. If one staple died because it was depleted or went out of fashion or whatever, another was there to sustain economic development.

Less fortunate regions had only one staple to rely on: one resource that the region's, or colony's, people saw as their only resource. These regions were less fortunate because, considered this way, staples are (by definition) export staples – destined for sale elsewhere – and thus their ultimate value is established outside the region. Hence, for example, CRC's constant concern about markets, which were the perennial focus of innovation in a business in which as much as possible was held constant. Markets, especially if they did not reside in the metropole, posed an added difficulty in managing the staple as a source of economic growth (for the region) or profit (for the firm). For fish, as for other export staples, markets were (and are) unpredictable – uncontrollable to a large extent; prone to glut; fair game for other primary producers from other places; liable to collapse under new production technologies which may produce synthetic forms of the original staple (beet sugar, for example), or new transportation technologies (steam) which can alter access patterns for the seller of the good – and because they were (and are) in the international arena, they were (and are) subject to problems of currency exchange, the vicissitudes of politics and war, and the whims of changing fashion. Hence, if it had only one export staple, a region was particularly vulnerable. If, as in Gaspé, that staple was controlled by one major firm whose purpose was profit, not regional development, the problem was compounded.

As it stood after the contributions of the Innis school of staple theory, then, a staple was:

- a resource;
- a resource for export;
- a resource for export which was depended upon to create economic growth in the producing economy.

Implicit in this increasingly complex concept is decreasing neutrality and increasing management, and an increasing burden of responsibility for the staple as a generator of wealth at the level of the regional economy. Export-base theorists, recognizing this, began to try to come to grips with *how* staples became engines for growth, and under which circumstances they were or were not successful.[4] In the process, diagnostic features of success and failure were identified. Baldwin and North[5] identified the technological conditions of the production function (the particular mix of land, labour, and capital that a given technology would demand) as important; so were rate of saving and level of education. These latter two they tied to the kind of institutional framework within which a staple was exploited: often, particular kinds of factor mixes would encourage particular kinds of institutional set-ups. Thus, the small-family-farm conditions of New England or southern Ontario, with their middle-income settlers, tended to promote evenly distributed incomes, an egalitarian society, and ultimate industrial development for an expanding domestic market via import substitution. But the cheap, low-grade labour structure of plantation economies promoted skewed income distribution, entrenched élites, no domestic market expansion and hence luxury imports coupled with home production of necessities, and a stagnant staple-production economy (the "staple trap").

To this more precise (but still often impressionistic) picture, recent work, especially that of Watkins and Gilmour,[6] added the capacity to measure the potential for growth in an export-staple sector, using the technique of Hirschman's linkages (see chapter 5). This endowed the export staple with a set of linkages which arose from potential inputs into, and outputs from, the resource, and which laid the path along which regional development could occur around the staple base, with given technologies producing given linkage potential. In the traditional merchant fishery, the dry cure added little value at source, unlike the industry involved in the timber staple, which could develop considerable forward linkages, such as the wooden sailing vessel. Backward linkages in the form of transportation were minimal in the fishery, and no complex machinery was required to create fish-processing factories. Given the truck system, there was little or no final demand either, and so consumer industries for a domestic market would not develop, unlike the boot-and-shoe and clothing factories or the breweries of agricultural southern Ontario.

If, then, the present connotations of the word "staple" are combined, the list becomes rather formidable. A staple

- is a resource (one or more);
- is exported from a colony (or region);
- has linkages which, if they are good, will generate growth around the export base. This growth will involved import substitution; ultimately it will lead to diversification beyond the export base, and the multiplier mechanism that results will create a take-off into self-sustained growth.

Along the way, however, the value of institutional structures seems to have become blurred: Baldwin and North's emphasis on élite versus egalitarian societies is no longer clear, nor is the recognition that, once in place, élites can become entrenched. But this is an essential component of the approach, and one that needs expansion, not exclusion. So to the list must be added that

- a staple has a set of socio-economic conditions attached to it, probably influenced by the nature of the production function under the initial technological conditions at the time of creation of the export sector. The resultant society and its institutions can affect the encouragement (or otherwise) of linkage development, education, distribution of income, capacity to transform or willingness to so do ... and so on. In other words, the staple is now also associated with the state or the ruling class. It has acquired a set of social relations along with the economic relations.

This kind of complexity in developing theory is exciting, because it links up with previously disparate work; the concept begins to crystallize in its entirety. With the social connotations of the term "staple" recaptured and refined, the powerful insights of metropolis/hinterland, core/periphery and development/underdevelopment thinking can be employed, their concepts of power, class, and institutional control be brought to bear on the issue, and the wider international arena in which commodity trades operate be appreciated.[7]

This is the enormous advantage of the detailed case study. It clarifies interrelationships that all too often remain undetected at a more general level of analysis. The choice of the fish staple is particularly fortuitous, precisely because it has been viewed as a "failed" staple – and failure in some staple economies has not been examined so much as assumed in the Canadian literature to date. As a result, thinking has been incomplete, vital parts of the staple economy have been downgraded, the connections between various components have not always been clearly seen, the role of the metropole has been taken as given or forgotten rather than being identified, the complete approach has not been adequately tested, and so responsibility (so to speak) for failure has fallen on the staple itself. Thus, Gilles Paquet writes that "the input of the cod economy in Canadian development has been marginal in all senses of the word" and that "in the list of leading sectors ranked by degree of development stimulation, fisheries are right at the bottom."[8] But Paquet was using only forward, backward, and final-demand linkages as his criteria here, and this is not enough. By using the expanded concept of "staple" developed in this book, the problem of limited vision that bedevils other approaches can be solved. The various components of a staple trade can be interwoven, and the issues of development and underdevelopment can be addressed. The relationship between the two becomes clear.

As with other staples, development of the fish-export staple sector started with a firm (in Gaspé) or series of firms (in the Atlantic region). Such firms were often, but not always, supported by the state in the form of metropolitan encouragement of the colonial development of a staple which was needed *by the metropole*. In the case of Jersey, the metropolitan end of the trade was very complex indeed, due to the peripheral nature of Jersey to the larger United Kingdom metropole. Nonetheless, just as Denmark needed fish from Iceland, or England needed cotton from the southern United States or timber from New Brunswick, so the West Country needed its Newfoundland-based cod fishery to replace the dying wool industry, and Jersey needed its New World fisheries to provide employment for a burgeoning population in the face of a limited metropolitan land-based economy.

In time, these staple firms became large enough to create a recognizable industry – a Jersey export sector, resident in the colony, which was a suitable generator of wealth and growth for the metropole first and foremost, and only incidentally for the colony. At this stage, the complex of firms involved in the colonial export sector became sufficiently important, in economic terms, that their individual profits – transferred back to the metropole and invested in other businesses or re-invested in the staple industry and associated spin-off industries – created a *metropolitan* multiplier out of the colonial staple trade. This was done, of course, not out of patriotism or conspiracy, but out of the perennial need of capitalists for certainty and security, a need that was deeply felt in the risky fish business with its difficult marketing situation. The combination of success and risk led to increasing institutional support, which is seen in this case study in the formation of the Jersey Chamber of Commerce and its growing power and support in the States of Jersey, in the Chamber's London patronage network, in the development of commission businesses in London and the link to the aristocracy, in the creation of Jersey banks out of fishing partners, and in the complex of tariffs, bounties, and facilitating legislation of various kinds that secured profitability, protection, and further investment. Metropolis/hinterland theorists and writers on merchant capitalism have been speaking of these things for a very long time.[9]

If the staple flagged, faltered, or went under, for any of the reasons already considered (poor markets, changing technology, war, fashion, resource depletion), then the institutional support of the metropole might be withdrawn or transferred, or might change form. After 1860 (in this instance), the British Parliament, operating under a laissez-faire ideology, was disinclined to support Jersey against French economic policy; after 1870, even the Chamber of Commerce itself grew cautious as it saw the weakening state of the cod trade and its associated carrying trade, as sail gave way to steam and a world commodity market evolved along lines antithetical to the old mercantilist cod fishery. By this time, too, the institutions of the new Canadian "core" (as opposed to the British "metropole") were growing in strength. Gaspé was coming to the notice of a Quebec government that had previously ignored it, and a wider regional economic potential was being created with the colonization roads. Perceptions of need were changing at the metropole and at the core, and the view of the Gaspé periphery changed accordingly.

In Canadian staple studies, however, which have been carried out mostly at and about the core, the institutional component has too often been ignored, or been seen as a competing approach. No allowances have been made for power. Instead, linkages have been the order of the day. This is because, when everything goes smoothly, it seems as though the system is mechanical, smooth-running, inevitable, and the nasty questions about theory do not get asked. This is the major problem, for example, with Gilmour and Paquet. One could coherently argue that the export-base approach has become caught in a perceptual staple trap. Staple theory has developed a rigour that the early Innisian approach did not have, but the export-base contribution has been applied without his breadth of vision. In making an operational model out of a view of the world, the emphasis on quantifiable mechanisms for growth has meant less emphasis on vital qualitative, or not easily quantified, components which gave the initial approach so much of its richness. To be specific, the export-base model of recent years has lost an international context and an awareness of the role of the metropole. The capacity of the Innisian approach to synthesize so many different strands of the story has been lost.

In 1960, Easterbrook said of the staple that it could "be viewed as a tool of analysis which enables study of total situations in terms of resources, technology and markets and the institutions, economic, political and social in which these are embedded … [it] leads to consideration of the institutions which emerge with its exploitation; the interaction of the staple production and this institutional complex; the outcome of this interaction and its consequences for later stages of development."[10] In 1972, Gilmour claimed that his "idealized theoretical industrialization of a region growing through the impact of its successful export sector upon the rest of the economy" – a very constrained interpretation of the staple approach – was "very general and imprecise," and could not hope to be otherwise if it were to "encompass all the eventualities and circumstances, social and economic, that made the development of real regions unique in character."[11]

In fact, a tension in thinking has developed around the staple approach: a feeling on the one hand that it must be rich and multidimensional, and on the other that the result of this is a serious loss of rigour. No wonder that one economic historian has seen it, in its guise of a commodity approach, as too difficult to handle.[12] Yet this is merely a staging post in the development of staples thinking, not a dead end. The approach has evolved, through intuition and vision, to the formulation of a limited operational model. That model has been tested empirically for the successful case of southern Ontario. This book has tested it for the so-called failed case of Gaspé, and that test has been useful because it has brought back into focus the critical power relationships that helped to structure this particular staple trade (institutional structures, constitutional manipulations, the role of élites, the motivations behind merchant capital) and it has pointed to the need to remember the international context of staple trades. In terms of rigorous methodology, it has highlighted the necessity of pursuing the analysis of linkage formation, not only in the newly settled region but also "back home," across the ocean in the metropole, something not previously done. This forestalls the old mind-set that allowed staples to be ranked merely by degree of development potential as measured in terms of local linkage effects: it should no longer be permissible to forget that linkage effects can occur elsewhere, and thus be not so much poor or nonexistent as invisible to the myopic Canadian perspective.

Baldwin, North, or Innis would not have had such blinkered vision, but Innis was essentially descriptive (albeit magnificently so), and Baldwin and North were side-tracked by their view of income distribution as critical. This is not to say that income distribution is unimportant. It is vital. But it is to say that a too-narrow concentration on income distribution misses the *theft* of linkages that is a major finding of this Gaspé case study. Such theft is not peculiar either to Gaspé or to the fish staple – the textile industries of England and New England were, after all, a stolen linkage from the cotton staple of the deep south of the United States. North and Baldwin, however, thought of uneven income distribution in terms of home production juxtaposed against luxury imports – pianos, fine furniture, fancy clothes. They did not consider that it might sometimes have been attributable to stolen linkage potential, which would result in the inability of a staple region to accumulate capital and thus create higher levels of final demand, which might ultimately lead to import substitution.

A detailed look at the Gaspé fish trade shows how such stolen linkage potential can rebound to the benefit of the metropole and the deprivation of the colony. It was not only (or even) luxury items that came to the Gaspé coast from Jersey. Potatoes and barrels were imported, as were fishermen's sweaters – not only those made in Jersey manufactories but also those knitted in Jersey homes, so that even cottage industry at the metropole benefitted. This was not a deliberate attempt to stunt indigenous development, although that was the result, but the returns from this strategy – to the firm in terms of profit and to the metropole in terms of multiplier effects – promoted development in Jersey at the expense of Gaspé. The same was true of shipbuilding, although the situation was more complex. The Jersey-built fleet started to grow only when an additional advantage began to accrue to the metropole with the repeal of the preferential duties on colonial timber, and when the developing cod trade generated a widening trade in the kinds of exchange commodities that allowed Jersey to purchase (with fish) the needed inputs for a domestic shipbuilding industry. Only then could the island develop a carrying-trade industry which gave it a new economy based on invisibles. What these examples reveal is a mercantile removal of cod-trade linkages out of Gaspé and metropolitan skill in benefitting from them. It does not speak to any inherent colonial inability to generate them. Pianos were certainly, as CRC objected to its Jersey agent on the Gaspé coast, a luxury item. Net and twine were not.

And so there has arisen a peculiarly Canadian confusion about what was originally Canada's contribution to economic theory. The impression has been created that there are "good" and "bad" staples, that these staples somehow stop at a regional boundary, and thus that there are "successful" and "failed" regional economies. This is a psychological staple trap and one which must be avoided at all costs, for it carries with it the burden of failure, inadequacy, frustration, and hopelessness. The Jersey-Gaspé cod trade brought people to Gaspé and wealth to Jersey. It brought development to the metropole and left a legacy of underdevelopment in the colony. But that underdevelopment was not a necessary consequence of the fishery *per se*. It was in large part a consequence of the manner in which metropolitan merchants operated a colonial staple trade using strategies which secured their business but, in the process, warped regional development. The golden age, in other words, was not so golden: it won its laurels at the metropole by sowing the seeds of future insecurity in the colony. This is important, because it implies that the historical yardstick so often used to assess the extent of present-day failure in the Atlantic Canadian inshore fishing economy is unrealistic and based on an inflated sense of past achievement. What the nature of that earlier achievement really was, what went wrong, and how, has to be clearly understood if it is to be a point of reference in any search for development potential in the future. This book has explored the roots of past "success" in the region's inshore fishery, and has specifically identified the manner in which it contributed to present problems. It is not only more encouraging, but more accurate to think of that past as a harsh reality with which to grapple, rather than as a demoralizing mythology which offers no solutions but instead haunts the regional psyche with a spectre of past glories now lost. Faced with a battle yet to be won, rather than a defeat to be endured, the people of Atlantic Canada may have a better chance of solving the dilemma of regional poverty.

Notes

INTRODUCTION

1 Innis, *The Cod Fisheries*.
2 See, for example, Alexander, "Newfoundland's Traditional Economy";
Alexander, *The Decay of Trade*; Balcom, "Production and Marketing";
Brière, "Le trafic terre-neuvier"; Chang, "Newfoundland in Transition"; de la Morandière, *Histoire de la pêche française*; Head, *Eighteenth
Century Newfoundland*; Matthews, "A History"; Ryan, "The Newfoundland Cod Fishery"; Turgeon, "Pour une histoire de la pêche." There is
also a small staples literature on Atlantic Canada, among the most recent proponents of which are, for example, McCann, "Staples and the
New Industrialism," and Wynn, *Timber Colony*. There is no historical
literature, however, on the Atlantic fisheries as a staple trade.
3 There are two works to date that deal with this fishery: Lee, *The Robins in Gaspé*, which is a biographical history, rather than an economic
history of the firm or the trade, and Lepage, "Le capitalisme marchand." There is also a work by Samson, *Fishermen and Merchants*, that
is concerned with the relationship between the firm of Hyman's of
Gaspé and its fishermen, but which does point to the dominance of
Jersey firms, particularly Charles Robin and Company, on the coast.
4 Price, "The Transatlantic Economy," 22.
5 See note 3. See also Chaussade, *La Pêche*, which is a general study of
fisheries in the region, including brief discussions of various technologies, various species, and the "golden age" of fish merchants (212–20),
and which also is concerned in a general way with regional
underdevelopment.
6 Jersey dominance is generally recognized in government reports
throughout the century and can be statistically verified by the 1870s,
when truly comparable data first become available. In 1874, for exam-

ple, those counties of Nova Scotia, New Brunswick, and maritime Quebec that were dominated by Jersey firms accounted for 31% of the region's cod production by value and 39% of all fishermen employed in any fishery of the region; see *Government of Canada, Sessional Papers, Appendix: Report on the Canadian Fisheries, 1874,* for data on production and unit export values.

7 McCusker and Menard, *The Economy of British America*; see especially p. 27.

8 Baldwin, "Patterns of Development"; Gilmour, *The Spatial Evolution*; North, *The Economic Growth*, especially Part 1; Watkins, "A Staple Theory."

9 Innis, *The Cod Fisheries*; Innis, *The Fur Trade*.

10 Price, "The Transatlantic Economy," 24.

11 For example, in Canadian studies see Acheson, "The Great Merchant and Economic Development"; Alexander, "Economic Growth"; Drache, "Rediscovering Canadian Political Economy"; Frank, "The Cape Breton Coal Industry"; McCallum, *Unequal Beginnings*; McCann, "Staples and the New Industrialism"; Naylor, "The History of Domestic and Foreign Capital in Canada"; Pentland, *Labour and Capital in Canada*. For the wider development literature see, for example, Brookfield, *Interdependent Development*; Brookfield, *Colonialism, Development and Independence*; Furtado, *Economic Development of Latin America*; Frank, "The Development of Underdevelopment" and several other writings, especially *Capitalism and Underdevelopment in Latin America*; and Girvan, "The development of dependency economics."

12 Head, *Eighteenth Century Newfoundland*; Innis, *The Cod Fisheries*; Matthews, "A History"; Ryan, "The Newfoundland Cod Fishery."

CHAPTER 1

1 *The Guernsey and Jersey Magazine,* 5 (1837): 306, an anonymous article entitled "The Commerce of Jersey." The original is "L'industrie d'une nation n'est pas bornée par l'étendue de son territoire, mais bien par celle de ces capitaux." (Author's translation.)

2 Information on the early history of Jersey is drawn mainly from Balleine, *A History*. For a detailed discussion of medieval land tenure, see deGruchy, *Medieval Land Tenures in Jersey*; for constitutional complexities, of which there are many, see Heyting, *The Constitutional Relationship*.

3 Balleine, *A History*, 94.

4 Ibid., 130.

5 For a detailed technical discussion of the highly complex system of feudalism and land tenure that was extant in Jersey see deGruchy,

Medieval Land Tenures in Jersey; for rents see, for example, 42. For the wider context, see Pollock and Maitland, *The History of English Law,* especially Book I, chapters 3 and 4, and Book II, chapter 1.

6 Balleine, *A History,* 130. See also the collection of papers of the Fiott family lodged with the Société Jersiaise and hereafter cited as the "Fiott Papers." A letter of April of 1792 shows John Fiott of London still exporting wool to Jersey and re-importing it to London in the form of hose for sale.

7 Poigndestre, *Caesarea or a Discourse on the Island of Jersey,* quoted in deGruchy, *Medieval Land Tenures,* 165.

8 DeGruchy, *Medieval Land Tenures,* 166.

9 Balleine, *A History,* 129–69. "Newfoundland" does not mean only the island of Newfoundland. Usually referred to in French as "Terre Neuve," the expression meant all the cod-fishing waters off the coast of eastern British North America.

10 See for example, the Janvrin connection with Southampton in Syvret, "Valpy Dit Janvrin," and Matthews, "A History," 70.

11 DeGruchy, *Medieval Land Tenures,* 129–30.

12 See Heyting, *The Constitutional Relationship,* for the intricacies of the local Jersey parliamentary system.

13 See Bloch, *"Les rois thaumaturges,"* for the definitive account of the King's touch.

14 Balleine, *A History,* 225.

15 Ibid., 236. The governor of Cherbourg commented: "Bons voisins pendant la paix, liés même assez étroitement par la contrabande, qui les enrichit, avec les habitans de la côte de Normandie et de Bretagne, qui les avoisinent, ils deviennent des ennemis très-dangereux dès que la guerre se declare, ou plutôt, ils sont toujours en état de guerre, tantôte contre les douanniers des deux royaumes, tantôt contre la marine marchande Francaise." [Good neighbours in peace, bound firmly to their neighbours of the Breton and Normandy coasts by the contraband which makes them rich, they become dangerous enemies in wartime – or, rather, it is always war: against the customs and excise of two kingdoms and the French merchant marine.] *The Guernsey and Jersey Magazine,* 309. See also Falle, *An Account of the Island of Jersey,* 382–3.

16 See Balleine, *A History,* 230 ff, who reports that Jersey was shipping in more tobacco than the island could consume. See also Syvret, "Valpy dit Janvrin," who details the Janvrin trade with America, especially New England.

17 *The Guernsey and Jersey Magazine,* 308. But these sets of figures seem exaggerated: 1771 was a particularly good year in the fishery, and 1,500 men seems excessive – it would entail eighty men to a vessel.

18 See Wright, *The Atlantic Frontier*, 195–6 and 260–84. For more detail, see Andrews, *The Colonial Period*, especially Volume III, 55–7 and 246. East New Jersey was sold for £3,400 and a peppercorn of quitrent; but in the Carolinas, Lord Carteret refused to sell his share for £2,500 and continued to try to claim an eighth of the quitrents.

19 For the history of the cod trade in Newfoundland, see Matthews, "A History." Matthews makes this point abundantly clear throughout his study. For more restricted work on the area, see Cell, *English Enterprise*; Head, *Eighteenth Century Newfoundland*; Matthews, *Lectures*; Quinn, "Newfoundland in the Consciousness of Europe"; and the papers by Barkham, Cell, and Lahey in Story (ed.), *Early European Settlement*. See also Mannion (ed.), *The Peopling of Newfoundland*.

20 Matthews, *Lectures*, 60–70, 83–8.

21 Matthews, "A History," 44–7, 58–9.

22 For a discussion of the limits of the French Shore, see Thornton, "Demographic and Mercantile Bases."

23 Head, *Eighteenth Century Newfoundland*, 54–6; Matthews, "A History," 156–259.

24 *Guernsey and Jersey Magazine*, 308.

25 Ibid.

26 "Name Files," Maritime History Archive, Memorial University of Newfoundland. These are a collection of raw historical data culled from a wide variety of sources by the late Keith Matthews, who generously gave me access to his files, as well as the benefit of his advice on this period of Newfoundland history.

27 Head, *Eighteenth Century Newfoundland*, chapter 7; Matthews, "A History," 378–428.

28 Head, *Eighteenth Century Newfoundland*, 166.

29 See Heyting, *The Constitutional Relationship*, Appendix, 160.

30 Minute Books of the Jersey Chamber of Commerce, November 1860.

31 Matthews, "A History," 18. One such occasion occurred after 1660, when some West Country merchants, short of "men, money and fish," sought the exclusion of Jersey, Irish, Scottish, and American ships from Newfoundland through a rigid application of the Navigation Acts; see Matthews, "A History," 189.

32 For a detailed contemporary account, see the *Journal de Daniel Messervy, 1769–1772*. The incident is in line with deGruchy's contention that there were seigneurial efforts to reinforce declining feudal structures in Jersey.

33 The second was Glasgow, under similar conditions of trade expansion outside the London commercial and political metropole.

34 The structure and function of the Chamber will be examined in detail

in chapter 3. See also Ommer, "A Peculiar and Immediate Dependency," for an extended discussion of the formation and function of this, and other, chambers of commerce.

35 See Smith, *An Inquiry*. Smith was contemporary with the rise of the Jersey Chamber of Commerce, and his work contains numerous passages on the tendency of merchants to "widen the market and to narrow the competition" (1, 267) and to "seldom meet together, even for merriment and diversion, but the conversation ends in a conspiracy against the publick, or in some connivance to raise prices" (1, 145). See also 1, 462, and many other comments.

36 "Rules of the Chamber," Minute Book 1, 1768.

37 Minute Book 1, 8 March 1768.

38 Ibid.

39 See "Harbour Grace Clearances, 1776–94," GN 11, Provincial Archives of Newfoundland. See also Ommer, "Peculiar and Immediate Dependency," tables 2 and 3.

40 Minute Book 1, 25 May 1772.

41 Matthews, "A History," 398, 428.

42 For a precise definition of this term, see Easterbrook and Aitken, *Canadian Economic History*, 53–5. Bye-boatkeeping entailed prosecuting the fishery inshore from small boats. This was done by fishermen hired in England who operated from temporary bases on shore and sold their fish to whoever would purchase it.

43 Matthews, "A History," 480–500.

44 Minute Book 2, 14 December 1785.

45 Matthews, "A History," 499–500.

46 Ibid., 495–6.

47 Name Files. See especially a petition from DeQuetteville to Governor Keates, 1815, in which he claims to have first established at Blanc Sablon, with two ships, in 1784.

48 BT 5/2.199 Board of Trade Enquiry, Buchanon's evidence, 10 January 1786; cited in Matthews, "A History," 398.

CHAPTER 2

1 Wade, "The Loyalists and the Acadians," 7–9.

2 *Report on the Canadian Archives*, 30 ff.

3 Lee, *The Robins in Gaspé*, especially chapters 1–3.

4 "Journal of Charles Robin," not paginated.

5 Chambers, *The Fisheries*, 111; Innis, *The Cod Fisheries*, 192; Saunders, *Jersey*, 213.

6 Innis, *The Cod Fisheries*, 193.

7 *Census of Canada*, 1931 (recap., vol. 1) 133–53.

8 *Report on the Canadian Archives*, 30.

9 Ibid.

10 Chambers, *The Fisheries*, 114.

11 *Report on the Canadian Archives*, 31. See 3 March 1773 and 14 June 1774 for the correspondence involved.

12 *Report on the Canadian Archives*, 30.

13 Ibid., 33.

14 Lee, *The Robins in Gaspé*, 37.

15 Charles Robin to Philip Robin, 13 May 1801, CRC Letterbooks.

16 Innis, *The Cod Fisheries*, 278, citing the *Quebec Gazette*, 30 April 1788, and 8 May 1788.

17 Innis, *The Cod Fisheries*, 279.

18 Charles Robin to Fiott deGruchy, 6 September 1792.

19 Charles Robin, quoted in Saunders, *Jersey*, 214.

20 The classic statement on this is Gordon, "The Economic Theory," 130–2, but it is inadequate to describe the dynamics of the industry under merchant capital.

21 Wynn, *Timber Colony*. The 200-mile limit, the first effective "fence" round Canadian fish, did not come until 1978. See Ommer, "All the Fish of the Post."

22 There is a parallel here with the tobacco trade as operated by Scottish merchants in the Piedmont region of colonial America. They recognized, as did Robin, that very small producers operating independently cannot make much profit, but that substantial profits could accrue to whoever could devise a way of aggregating individual returns. See Egnal and Ernst, "An Economic Interpretation of the American Revolution," 25.

23 Ryan, *Fish Out of Water*, especially chapter 3.

24 Indeed, this problem of expansion, overproduction, glut, and consequent depression has been endemic in the fish trade throughout its history. Even today it remains a problem: see the Kirby Task Force's report on the over-expansion and glut problems of the Canadian fishery after the imposition of the 200-mile limit, and government's attempt to deal with them: *Navigating Troubled Waters*.

25 Crown land, which was most of the coast, was free and given to those who were awarded such grants by Parliament. Land bought in Cape Breton by DeCarteret and LeVescomte in 1844 cost £40 Halifax currency for sixty acres when purchased from the General Mining Association; the same firm paid £115 for thirty-two acres of prime shore land at D'Escousse in 1834. DeCarteret and LeVescomte Papers.

26 Lee, *The Robins in Gaspé*, 35.

27 And even after the firm was taken over by Halifax interests in 1910. Jerseymen were, in fact, still being employed as clerks and managers as late as 1924. Interview with Arthur LeGros, late manager of the firm at Paspébiac, September 1977.

28 Minute Books of the Jersey Chamber of Commerce, especially Books 1 and 2.

29 CRC Letterbooks, January–March 1840, give an excellent account of winter activities at Paspébiac. See chapter 4.

30 See, for example, Baldwin, "Patterns of Development," 68–75.

31 Glenalladale Papers. The rent for this property over the 2,996 years follows in the original.

32 Baldwin, "Patterns of Development."

33 Ommer, "Scots Kinship," especially chapter 3.

34 In the case, for example, of a man whom a firm was particularly anxious to hire, and who was therefore in a position to strike a bargain with the merchant.

35 I am indebted to Gordon Handcock, of the Geography Department of Memorial University of Newfoundland, for the information given here on West Country practices.

36 Keith Matthews, personal communication, 11 October 1978.

37 See, for example, the Perrée Papers, in which there are several similar examples of indentures. I am indebted to Henry Perrée, of Jersey, for permission to study his account book of the Perrée firm, now lodged in the Société Jersiaise, along with a set of letters written by the firm, which are hereafter cited as the Perrée Papers.

38 See, for example, Raulin Robin and Moses Gibaut to Paspébiac agent, 5 September 1876: "*Seaflower* sailed this morning ... Mr. Frcs. [Francis] [Francis] Gibaut ... on board ... paid passage." Papers in the possession of the family of the late Arthur LeGros of Paspébiac. (Hereafter referred to as "LeGros Papers.") These are a collection of CRC papers privately held by the LeGros family, containing documents and letters of CRC (later Robin, Collas and Company, and then Robin, Jones and Whitman).

39 See, for example, Paspébiac agent to CRC (Jersey), 3 November 1815: Captain Le Gresley, he noted, "bought part of Mr. David's premises at Newport and hired people to build a stage and a flake." LeGros Papers.

40 Paspébiac agent to James Robin (Jersey), 5 May 1840, noting one Charles Remon as having been missing.

41 Le Feuvre, *Jèrri Jadis*, 121. [The Coast! A country that has a fascination and a quite special bond of feeling for an old Jerseyman like me ... A country that has seen ... the toil, the sorrows – and the joys also – of

daily life for so many Jerseyfolk, born in Jersey but moved to the coast, to its borders and the foot of its mountains.]

42 For a detailed description of merchant/labour relations in Gaspé, see Samson, *Fishermen and Merchants*.

43 Innis, *The Cod Fisheries*, 155, observed that "with the rise of the truck system ... went the growth of a resident population, the importation and distribution of supplies, the purchase of fish, and their collection and sale ..."

44 "Engagement of Jules Carron, Percé," dated Percé, 17 December 1873. LeGros Papers.

45 United Province of Canada, *Sessional Papers*, Report of Pierre Fortin on the Gulf of St. Lawrence Fisheries, 1864, 20. (Most of the reports used in this study were those of Pierre Fortin, and are cited hereafter as the *Fortin Report*, along with the year of the report.) [The fishermen are taken to the fishing stations at the expense of the company that hires them; they have for their use a good boat, fully equipped ... and they receive for each 100 cod that they weigh at the stage the sum of 5/6d, half in goods and provision ... But they feed themselves at their own expense, and if the fishery is poor their provisions account uses up most of their profit and they often return to the Magdalen Islands empty-handed.]

46 Charles Robin to John Henderson, of Miramichi, 31 May 1788.

47 Letter from Jersey to Paspébiac headquarters, 14 March 1854. LeGros Papers.

48 Charles Robin to Fiott deGruchy, of London, 29 August 1796.

49 See, for example, Antler, "Colonial Exploitation"; Faris, *Cat Harbour*; Innis, *The Cod Fisheries*; Sider, "Christmas Mumming."

50 For a detailed discussion of truck and its various functions, see Ommer, "The Truck System."

51 Paspébiac headquarters to James Robin in Jersey, 10 December 1845.

52 See, for example, the letter cited above.

53 O'Hara to Haldimand, 12 September 1785. *Report on the Canadian Archives*.

54 Pentland, "The Role of Capital," 461. See also Hilton, *The Truck System*, which remains the definitive work on the truck system.

55 Letter from Paspébiac headquarters to M. Thomas Proux, Québec, 9 August 1825; and to M. Boullet, of St. Thomas, Quebec, 1 May 1844. LeGros Papers. [If the salters Boulet, Nicholle, and J. Bernaiche wish to work for us again you can offer them one or two more piastres per month than what I specified, and likewise for good splitters who have already been with us and whom you know well.] See also Mountain, *Visit to the Gaspé Coast*, 26.

56 Although they exercised their usual caution in payment, and kept it to a minimum. See letter to M. Boullet (note 55, above), in which CRC attempted to defray rising transportation costs with a cutback in wages.

57 Letter of E. Mabé, shipbuilder, "Corner the Beach" (Malbaie), to Elias leBas, merchant of Jersey, 14 November 1844. Perrée Papers.

58 Samson, *Fishermen and Merchants*, 23.

59 Innis, *The Cod Fisheries*, 277–8.

60 Charles Robin, quoted in Saunders, *Jersey*, 214.

61 Ibid.

62 Charles Robin to CRC (Jersey), 22 November 1794.

63 Charles Robin to Fiott deGruchy, 1 December 1795.

64 Charles Robin to Fiott deGruchy, 28 May 1796.

65 Ibid.

66 Charles Robin to Fiott deGruchy, 11 November 1796.

67 Ibid.

68 The war undoubtedly affected this firm, but the evidence is lost.

69 Charles Robin to Fiott deGruchy, 12 July 1797.

70 Ibid.

71 Charles Robin to Thomas Amory (Boston), 7 November 1797.

72 Charles Robin to CRC (Jersey), 16 November 1797.

73 Charles Robin to Burns, Woolsey (Quebec), 1 March 1798.

74 Charles Robin to Burns, Woolsey (Quebec), 1 August 1798.

75 Charles Robin to P. and H. LeMesurier (the firm's new London commission agents), 24 October 1798. He had avoided convoys up to this point because large numbers of ships arriving together at market created gluts and, consequently, low prices.

76 Charles Robin to Burns, Woolsey (Quebec), 13 May 1799. The *St. Lawrence* was captured again – near Vianna, by "two Spanish rowboats" – but was saved by an armed boat which re-took her.

77 Charles Robin to Burns and Woolsey (Quebec), 13 May and 25 June 1799. The Admiralty sentence on her at Vianna was in favour of CRC, but had to be confirmed in Lisbon. Hence, she would be unable to leave Portugal before August, and would then be a prime target for privateers.

78 Charles Robin to P. and H. LeMesurier, 27 November 1799. The letter includes a highly personal view of the French in the war, which leaves no doubt as to Robin's personal allegiance to Britain, and to his hopes and arguments for a treaty favourable to the better protection of the fisheries from French competition. "Let the English bestow their generosity and favour on the brute creation," he wrote, "there they will find acknowledgement and gratitude ... Let us never forget

that they [the French] are a set of disappointed Tigers ... therefore let
France be a Kingdom or a Republic. The D...l trust them, but an En-
glishman should never do it."

79 Charles Robin to Philip Robin, 22 July 1800.
80 Ibid.
81 When this had happened to Robin, Pipon and Company in the American War of Independence, Robin had been forced to abandon Paspébiac.
82 Charles Robin to Philip Robin, 13 May 1801.
83 Ibid.
84 Charles Robin to Philip Robin, 27 October 1801.
85 Charles Robin to Burns, Woolsey (Quebec), 4 January 1802. As a merchant, he did not like it: "All the particulars I know that we give up
   all our conquests except Ceylon, Trinidad and Tobago, which in my
   opinion is dearly bought."
86 Charles Robin to Philip Robin, 22 July 1800.
87 Charles Robin to Philip Robin, 31 May 1802.

CHAPTER 3

 1 Matthews, "Pipon Family," 1.
 2 Syvret, "Valpy dit Janvrin," 27.
 3 Minute Books of the Chamber of Commerce.
 4 Syvret, "Valpy dit Janvrin," 31.
 5 Ibid.
 6 Journals of the House of Assembly, Newfoundland, 1859, 373. See also
   Thornton, "Demographic and Mercantile Bases," 181, fn 15.
 7 Thornton, "Demographic and Mercantile Bases," 177.
 8 Name Files: Slade file.
 9 Thornton, "Demographic and Mercantile Bases," 177.
10 See DeQuetteville's comments in the Chamber of Commerce Minute
   Books, 9 September 1863. See also Innis, The Cod Fisheries, 412–13.
11 Thornton, "Demographic and Mercantile Bases," 177. The process was
   not confined to the Strait of Belle Isle. Baine Johnstone, for example,
   took over Slade's operation in Battle Harbour. See Name Files: Slade
   file.
12 See partnership breakdowns in the Jersey Ship Registers, 1803–30.
13 Fiott Papers, letter of 1771. The Fiott Papers are a collection of personal and business papers of the family, all unfortunately fragmentary. They are held in the Société Jersiaise.
14 Fiott Papers, remnants of a journal of 1772.
15 Name Files. His father bought the seigneurie.

16 Fiott Papers. He wrote to his father in 1774 that the "gentlemen of St. Aubin's" (Robins, Pipons et al.) would be with them.

17 Fiott Papers, letter of 2 May 1780.

18 Fiott Papers; a set of letters dated 1781–85.

19 Letter of John Fiott in London to Sir William Lee, 29 March 1783.

20 Ibid.

21 Ibid.

22 John Fiott to William Lee (London), 31 December 1781.

23 As the Jersey economy diversified, the members grew in number and broadened in the scope of their interests. However, the prime importance of the cod trade until 1860–70 is clear.

24 Minute Books, January 1796.

25 Minute Books, 20 April 1805.

26 Minute Books, 23 May 1786.

27 Minute Books, November 1786.

28 Minute Books, 14 April 1806. Copy of a letter (no. 1) sent to the Honourable Board. See also no. 13 to J.T. Swanson.

29 Minute Books, Letter no. 28, 19 August 1806.

30 Minute Books, 16 March 1816 and 23 March 1816.

31 Ibid. The restrictions placed on Jersey in the proceeding years were part of a general British protectionist policy of rigorous taxation. See Imlah, *Economic Elements*, 166, where he states that "the compelling problem was revenue." Attitudes leaning toward free trade came to the forefront only after 1824, and in the 1830s Parliament started to work toward such legislation.

32 Minute Books, 23 March 1835: Thomas Luck to Th. LeBreton.

33 Minute Books, 31 March 1835: Thomas Freemantle to Th. LeBreton.

34 Ibid. Charles Le Quesne to the Spanish Consul General in London (Señor Don José Maria Barrero), 31 December 1841. This would appear to be a development in the market strategy of Jersey that has not been recognized up to now. The ability to stockpile fish until market demand occurred (six months later) would be of considerable benefit to the cod trade, and hence the proposed tariff was a serious threat to that strategy.

35 Ibid.

36 Minute Books: Philip DeQuetteville to the president of the Board of Trade, December 1841. It is worth noting that DeQuetteville had good reason for regarding the cod trade with "peculiar interest," since he was one of the principal cod merchants, with extensive premises in Labrador.

37 Ibid.

38 Minute Books, March 1842.

39 Minute Books, 12 June 1860. See the discussion earlier in this chapter.

40 Minute Books, 25 July 1860.

41 See Imlah, *Economic Elements*, 17.

42 Minute Books: M. Hobiers of St Malo to Abraham deGruchy, 30 October 1860. [It becomes difficult to purchase your produce because the Channel Islands are excluded from the commercial treaty between France and England and our customs and excise are becoming more strict.]

43 Minute Books, November 1860.

44 For the importance of this situation for Newfoundland and Labrador, and the fish-ship tradition, see Innis, *The Cod Fisheries*, 412–13.

45 Minute Books, 9 September 1863.

46 Minute Books: Frederick Rogers (Downing Street) to Joshua LeBailly, 5 December 1863.

47 Innis, *The Cod Fisheries*, 412, notes that this same legislation resulted in the displacement of many old English firms by Newfoundland firms on the Labrador. See also Thornton, "Demographic and Mercantile Bases," 177.

48 Minute Books, 26 January 1786. See also 2 March 1787 and 7 March 1787.

49 Bailyn, *The New England Merchants*, especially 189–90.

50 Beamish, Hillier, and Johnstone, *Mansions and Merchants*.

CHAPTER 4

1 Cook, *The Headguts and Soundbone Dance*, Act 2, 46.

2 Robin, Jones and Whitman Papers, MG 28 111 18, Letterbooks, *passim*; Fortin Report, *passim*.

3 Figure 10 should be compared to figures 23 and 24, because the cod-trade shipping network shown there is the spatial geometry of the system described here.

4 See, for example, Templeman, *Bulletin #154*. See also Jean, "Seasonal Distribution of Cod," 429–60. Templeman, 38–9, points out that cod migrate out of the Gulf of St Lawrence with the approach of winter, moving up the west side of the south coast of Newfoundland, where they stay in deep water "until after April, when they return along the West Coast of Newfoundland into the Gulf of St. Lawrence." They spend the winter "in deep warm water below the cold layer, often 100 miles or more from the coast ... After spawning, large numbers move ... to the coast, feeding on caplin ... Their availability is affected greatly by variations in temperature ... onshore and offshore winds ... intensity of light and spawning places of the caplin" (p. 2).

5 As early as 1611, the Church of St Brelade's held its spring Commu-

nion earlier than the other parishes, to accommodate the increasing numbers of its parishioners who were involved in the fishery. See Balleine, *A History*, 129.

6 Charles Robin (Paspébiac) to Philip Robin (Jersey), 16 November 1799. By November and December of each year, Charles was writing back to Philip with instructions and requests made in preparation for the following season. In later years, with agents rather than the owners on the (Gaspé) coast, instructions came from Jersey at the beginning of each season, in response to information and requests from Paspébiac.

7 Some stores, especially flour and meat, were acquired from Quebec (and later Halifax) by the Paspébiac headquarters.

8 Charles Robin to Thomson, Croft and Co. (Oporto), 11 August 1777.

9 Charles Robin to Fiott deGruchy (London), 11 August 1777. "Trying to dispatch *Bee* for Jersey by 25 September ... If she can reach Jersey safely she must be ... sent to Bilbao and afterwards to Lisbon for a cargo of salt." See also chapter 3, section on the Spanish tariff.

10 Charles Robin to Fiott de Gruchy, 20 November 1777.

11 After 1815, Jersey amassed stores increasingly through Liverpool, some ships operating a route from Liverpool to the Gulf of St Lawrence to markets, in a series of trading voyages in which Jersey was not visited for several years. Instead, men were sent from Jersey to Liverpool to join the cod-ship crews, and Jersey agents in Liverpool (such as Robin and King) managed the "fitting out" operations.

12 *The Guernsey and Jersey Magazine*, 311. "The vessels for the fishery prepare for the voyage in March and April; and some sail at the latter end of the former month, but many more during the latter."

13 Minute Books of the Jersey Chamber of Commerce, 7 March 1787.

14 *The Guernsey and Jersey Magazine*, 311.

15 Flour usually came from Quebec or Halifax, with only a little from Jersey against the chance of shortage. Biscuit, however, was a Jersey manufacture for the fishery and was always brought on the ships.

16 Charles Robin to G. Allsopp (Quebec), 2 August 1777.

17 Charles Robin to Jersey headquarters, 11 August 1777.

18 Arthur LeGros, personal communication, September, 1977. The smaller-grained (finer) salt provided a more even cure. The use of coarse salt required more care and skill to turn out a quality product.

19 Especially for planters, who preferred it because it was cheap, despite the poorer cure it provided.

20 Charles Robin to Philip Robin, 1 October 1800. Good labour, especially at the management level, was an ongoing concern. See letters even as late as 4 November 1876. LeGros Papers.

21 Paspébiac agent to M. Thomas Proux (Quebec), 8 September 1825. LeGros Papers. (See footnote 55, chapter 2.)

22 Charles Robin to Fiott deGruchy, 1 December 1795.

23 Charles Robin to Fiott deGruchy, 9 June 1795.

24 See United Province of Canada, *Sessional Papers*, Reports on the Fisheries, *passim*. See also Hubert, *Les Îles de La Madeleine et les Madelinôts*, especially 108–248.

25 "The *Hilton* must be dispatched early from hence to Lisbon with a cargo of old and Fall fish for sake of a full cargo of good salt." Charles Robin to Fiott deGruchy, 1 December 1795.

26 Ryan, *Fish Out of Water*, is the authority on markets in the Newfoundland fishery; there is no equivalent for the Maritimes.

27 Charles Robin to Thomson, Croft and Co. (Oporto), 11 August 1777.

28 Charles Robin to Messrs. Ventura Francisco Gomez & Berena, 16 October 1783. Robin set up CRC (as opposed to Robin, Pipon and Company) in 1783, hence the "old friends."

29 Charles Robin to William Gray, 29 August 1791.

30 As did most merchants in the cod trade, since the credit system forced them to accept payment in whatever form the planter could give it – often in commodities such as furs, staves, or salmon.

31 Charles Robin to Robin, Pipon and Company (Jersey), 8 November 1777. That year, he sold his moose skins and feathers to William Smith (see chapter 2) on the coast for £120 sterling, and sent his furs to London. The next year he explained the complications of marketing fur more completely: "Please to consider whether landing the Furrs at Falmouth will not be running too great a risk to have them carried to London, and a great additional expense owing to the premium. Couldn't it be managed to send them from Jersey to England?" – Charles Robin to Fiott deGruchy, 25 July 1778.

32 Charles Robin to G. Allsopp (Quebec), 11 September 1778. The furs of the previous year were: 1,200 beaver, 1,200 marten, 50 otter, 50 fox, 300 muskrat, 50 castoreum (beaver-gland extract). Charles Robin to Fiott deGruchy, 11 August 1877.

33 Charles Robin to John Urquhart, planter, 11 July 1777.

34 See, for example, the Fortin Reports, *passim*. One such cargo, on the *St Lawrence*, contained 2,060 quintals codfish, 60 tierces salmon, 11 barrels salmon (netted), 48 barrels salmon (speared), and 12 hogsheads of cod oil, all destined for Bilbao. Charles Robin to Gomez of Bilbao, 10 December 1785. In 1788, Robin agreed to take the salmon of J. Henderson of Miramichi, because it was of good quality and had sold the previous year at Leghorn, at 37/6 less 8/– freight.

35 See chapter 2. It could be difficult to market – see Charles Robin to Mr Amory (Boston), 21 May 1798: "*Hilton's* cargo of staves unsold at Lisbon" – because too high a price was asked.

36 Charles Robin to Newman and Land (Oporto), 22 July 1786.

37 Innis, *The Cod Fisheries*, 301 and footnote 38; Prowse, *A History of Newfoundland*, 403. Twenty-six hundred quintals were exported to Brazil in 1812; next, in 1814, was 2,049 quintals. Matthews, however, comments that *Lloyds Lists* show the first St John's vessels sailing to Pernambuco in 1809 – personal communication, 3 October 1978.

38 Innis, *The Cod Fisheries*, 209; Fay, "South American and Imperial Problems," 183–96.

39 Letter of 3 November 1835 to Jersey from the agent at Paspébiac.

40 *The Guernsey and Jersey Magazine*, 311.

41 A. LeGros, personal communication, September 1977. See also, for example, Head, *Eighteenth Century Newfoundland*, 16, 72, 140, 243, 249 (heavy "wet") and 3–5, 6, 16, 72, 74, 140, 141, 243 (light "dry"); Innis, *The Cod Fisheries*, 222, footnote 22. See also, for the modern situation, Alexander, *The Decay of Trade*, especially 45–6, 54, 92, 96, 130–1, 136–7. Note, as well, Charles Robin to Philip Robin, 31 May 1802: "If we can make up 300 qtls. [of old and fall fish] we'll put no new on board ... I think it would have answered better at Bilbao than anywhere else from its quality."

42 Charles Robin to Fiott deGruchy, 28 May 1796.

43 Charles Robin to Captain Jean Norman of the *Hope*, 8 November 1777.

44 Letter to John Fiott from his agents in Alicante, 10 December 1771. Fiott Papers.

45 Charles Robin to Jean Norman, 8 November 1777. ["It is always needed."]

46 Letter to John Fiott from his agents in Alicante, 18 January 1772. Fiott Papers.

47 *The Guernsey and Jersey Magazine*, 47.

48 C.W. Robin (Jersey) to Paspébiac agent, 29 April 1865. LeGros Papers.

49 See, for example, Jersey headquarters to Paspébiac, 9 May 1876, LeGros Papers: "*Ranger* to sail for Turks Island, thence to you. Take note she had 10 bags coffee on board."

50 See the advertisements in Jersey newspapers, cited in chapter 6.

51 *The Guernsey and Jersey Magazine*, 47.

52 James Robin and C.W. Robin to Paspébiac agent, 2 May 1860. LeGros Papers.

53 James Robin and C.W. Robin to Paspébiac agent, 12 December 1871.

54 James Robin and C.W. Robin to Paspébiac agent, 29 August 1876.

55 Head, *Eighteenth Century Newfoundland*, 249.

56 See, for example, the yearly instructions from Raulin Robin for CRC (Paspébiac), LeGros Papers.

57 C.W. Robin to CRC (Paspébiac), 8 November 1865. LeGros Papers.

58 Paspébiac agent to Raulin Robin (Naples), 16 September 1847.

59 The establishment of regular steamer services to Quebec, Boston, Halifax, and major market ports helped. The arrival of the telegraph made possible much faster communication. See, for example, Raulin Robin (Jersey) to CRC (Paspébiac), 8 August 1876: "Telegram from Rio ... *Robin* arrived 7 inst. Sold 25 $500 ... returns in ballast ..." LeGros Papers.

60 In later years, as communications improved and the steamer and telegraphy reduced market uncertainty by increasing and speeding up communications and information flow, the element of skill in market decision making decreased, the risk lessened, and the possibility of a "windfall" gain likewise decreased. Thus, the whole trade became more marginal for the merchant as it became more certain.

61 Charles Robin to Burns and Woolsey (Quebec), 1 August 1798.

62 Charles Robin to Philip Robin, 31 May 1802.

63 All quotations from Paspébiac agent to James Robin in Jersey, 30 (?) April 1840 (date unclear). The letters are unsigned, and were written by the chief agent at Paspébiac.

64 Paspébiac agent to James Robin, 5 May 1840.

65 Paspébiac agent to Creighton and Grassie (Halifax), 5 May 1840.

66 Paspébiac agent to Creighton and Grassie (Halifax), 19 May 1840.

67 Paspébiac agent to James Robin (Jersey), 19 May 1840.

68 Ibid. "We yesterday shipped per *Old Tom* 307 quintals of fish."

69 Paspébiac agent to James Robin, 25 May 1840. For hiring practices, see chapters 2 and 5. Skilled French-Canadian shore crews were hired through a Quebec agent, on a monthly basis.

70 Ibid. The problem was that this would lessen the amount of fish brought to CRC for purchase, and hence would diminish quantities sent to market.

71 Paspébiac agent to James Robin, 19 May 1840.

72 Paspébiac agent to James Robin, 25 May 1840. That is, not having been supplied with salt to cure their fish, they would bring it unsalted to CRC for sale at a lower price. The letter does not say why, but LeBoutillier Frères were new on the coast as a firm in their own right, and possibly not yet trusted by the local planters.

73 Paspébiac agent to James Robin, 20 May 1840. The firm built ships regularly, usually one per year. See CRC Letterbooks, *passim*; Fortin Report, *passim*; and, for changes over time, chapter 6.

74 Paspébiac agent to Creighton and Grassie (Halifax), 9 June 1840. LeBoutillier Frères, being newly formed, did not yet have a market-information network that could operate as efficiently as that of CRC. It appears from the letterbooks that, in this year at least, the firm's strategy was to send all ships to Jersey for market orders. Toward the end of the year, when the fall fishery failed, CRC had to resort to a more

evolved version of this strategy: sending its ships to market "central information points," such as Naples, there to receive orders from Jersey which awaited them in these central ports. Clearly, market decisions could make the difference between success and failure of the year's venture.

75 Ibid.

76 Ibid. Sometimes fish were bought "green" by CRC and then cured by the firm on its premises, to ensure a good-quality cure. A. LeGros, personal communication, September 1977.

77 Ibid. This is a good-quality cure.

78 Paspébiac agent to Creighton and Grassie (Halifax), 16 June 1840.

79 Paspébiac agent to Creighton and Grassie (Halifax), 16 June 1840.

80 Paspébiac agent to Messrs. Richard Bertram and Company (Civita Vecchia), 17 June 1840.

81 Paspébiac agent to James Robin, 29 June 1840.

82 Paspébiac agent to Creighton and Grassie (Halifax), 20 June 1840. The same letter indicated the arrival of the *Dit-on* with supplies, including seven puncheons of Berbice rum.

83 Paspébiac agent to Dobree, Maingay and Co., 4 August 1840. Captain Huelin was to get one hundred quintals, and his mate, George LeGros, fifteen quintals, "as privilege," which was a standard practice that helped to ensure an officer's loyalty when he was trading in the marketplace.

84 That is to say, was buying fish up and down the coast.

85 Paspébiac agent to Dobree, Maingay and Co., 4 August 1840.

86 Paspébiac agent to James Robin, 10 August 1840.

87 Paspébiac agent to Bertram and Co. (Civita Vecchia), 20 August 1840. They were instructed to refer to the Jersey "principal" (James Robin) for instructions as to ship and cargo.

88 Paspébiac agent to I.H. Gosset, 21 August 1840.

89 Paspébiac agent to James Robin, 1 September 1840.

90 Paspébiac agent to James Robin, 13 September 1840.

91 Paspébiac agent to James Robin, 20 September 1840.

92 Paspébiac agent to James Robin, 1 October 1840.

93 Paspébiac agent to Captain LeGros, 2 October 1840. "You merely call in Jersey to take orders ... You will not run in, but come to the roads and go onshore with your letters, to consult James Robin, Esq., taking care that none of your crew leave you there, the voyage not being ended..."

94 Paspébiac agent to Messrs. Hunt, Roope, Teage and Co. (Oporto), 3 October 1840, stating that they would "have advice of this shipment from our Principal in Jersey. Also his views for the procedure of the vessels..."

95 Paspébiac agent to Creighton and Grassie (Halifax), 10 October 1840.

96 Paspébiac agent to Burns, Symes and Co. (Quebec), 25 October 1840; and to Dobree, Maingay and Co. (Naples), 6 November 1840.

97 Paspébiac agent to James Robin, 7 November 1840.

98 Ibid. Haddock in the Baie des Chaleurs were increasing, and "in future must be kept entirely separate." The Mediterranean markets did not mind – "where they make no difficulty, as long as the quantity is not too great."

99 Paspébiac agent to William Cuthbert and Co. (Quebec), 27 November 1840; and to Messrs. Stewart (Dalhousie), 27 November 1840.

100 Charles Robin to Philip Robin, 22 June 1801.

101 Charles Robin to Philip Robin, 13 May 1801. See also Charles Robin to Philip Robin, 22 November 1794: "Deprived of all my letters, I could not form any ideas where to send the fish, it being late before I could get it all shipped and much of it having a dark cast, I thought the Southward was by no means a proper place to send it, and thus all is lost."

102 Paspébiac agent to James Robin, 7 November 1815. Supplies were given out in accordance with instructions received from Jersey at the beginning of the season. This edict seems odd: salt is a high-volume, low-value commodity, normally dispensed by the hogshead, with no need for such accurate measurement. Such caution in supplying salt is very unusual.

103 CRC (Jersey) to Paspébiac, 29 May 1860. LeGros Papers.

104 Paspébiac agent to Captain Huelin of the *Homely* at New York, 8 June 1847.

105 CRC (Jersey) to Paspébiac, 14 March 1854. LeGros Papers.

106 On 3 October 1855, the central London commission agents – DeLisle, Janvrin, deLisle – went bankrupt. CRC thereafter handled the "banking" business itself, from Jersey. For the Janvrin failure, see the *Shipping and Mercantile Gazette*, 4 October 1855. This firm's liabilities were £400,000, and it was described as "one of the oldest and most respectable houses in London." The cause was given as heavy financial involvement in Canada, principally advances made to "an extensive firm at Quebec" which had been in trouble in 1837, and was again in difficulty following "pressure in New York and Canada about two years ago." The full sum of the advances was estimated at £200,000, and little hope was held out that it would be retrieved.

107 In 1831, Philip Perrée transferred his property at Point St Peter, Gaspé, to John Perrée, Jr, who sold it to John Perrée, Sr. Two years later he transferred the property to his son, Francis, who sold it to the Gaspé agents J. and E. Collas in 1851. This was the firm that would take over CRC in the 1880s, when it failed. Jurat Perrée com-

ments that the family sold out of the business to go into a flourishing local potato trade with London (see chapter 7). Interview, November, 1976, St Helier. A more detailed picture of the cash flow would have been possible if the account books of the London agents had been available for analysis. However, an exhaustive search of all major London archives failed to produce records of this, or any, Jersey–London commission house, and no such records are known to exist in Jersey, other than the fragmentary Fiott Papers in the Société Jersiaise.

108 The Perrée firm is an excellent example of the opportunities that existed on the edges of the cod trade. It operated in a parasitical fashion off the larger firm, providing supplementary freight space for CRC and doing minimal fishing of its own.

109 Warrant no. 41. Entry for the brigantine *Adventure*, Custom House, Gaspé, 10 June 1845. Perrée Papers. There were no other customs declarations in the collection for that year.

110 DeLisle, Janvrin, DeLisle handled the Perrée account, as well as that of CRC. In the Perrée Account Book, the relevant policies are fos. 21 and 40, with fos. 6–12 and 56 as contra accounts.

111 Perrée Account Book, fos. 13–17 and 37; Contra fos. 18, 37, 38, and 43.

112 Ibid., fos. 1, 2, 33, and 34; contra fos. 4, 29–36.

113 Ibid., fos. 21 and 40; contra fos. 6–12 and 56.

114 Ibid., fos. 53 and 54; contra fo. 19.

115 Ibid., fo. 50: £60–1–6 in April 1848.

116 If the 1854 turnover at Quebec (£261–0–5), which invoiced the Gaspé establishments, is taken, and £60–1–6 is added for salt (from Dean and Mills of Liverpool), it would appear that the Gaspé establishment would cost about £321 per annum, that is, 23% of the profits of 1848.

117 Charles Robin to Thomson, Croft and Co. (Oporto), 11 August 1777.

118 Charles Robin to Foreman, Grassie and Co. (Halifax), 8 May 1797. "This catastrophe, although a great disappointment, will not in the least affect our Employ as we are perfectly secured."

119 One of several "bills of exchange" and "orders" sent to G. Allsopp (Quebec), 29 July 1783, as "draft bills to draw," so that Allsopp would be familiar with the format. LeGros Papers.

120 Letters from Alicante: 11, 18, and 21 January 1772. Fiott Papers.

121 Charles Robin to G. Allsopp (Quebec), 8 September 1777: "Please send our account current home in the Fall"; and Paspébiac agent to CRC (Jersey), 6 July 1852: "Messrs. Creighton and Grassie have drawn on Messrs. DeLisle and Co. for balance due them, say £732–14–1 sterling, 13 p.c. prom" (i.e., 13% promissory notes).

122 Charles Robin to Thomas Ainsley, H.M. Collector of Customs (Gaspé), 17 September 1794: "We pay yearly to Canadian fishermen about £1200."

123 Charles Robin to R. and P. Bruce (New York), 28 May 1796: a letter instructing them to remit the proceeds of goods through Burns and Woolsey (Quebec) in "good Bills on London."

124 Charles Robin to CRC (Jersey), 10 November 1795.

125 Charles Robin to Antonio del Campo, 27 August 1791.

126 Charles Robin to Philip Robin (Jersey), 20 November 1792. Such as had Fiott and Company: "an excessive scarcity of money ... has put it out of our power ... to procure any ready cash or specie to give your said Commander." Letter from Alicante, 7 December 1771. The end of the mercantilist philosophy in Britain ended the thirst for bullion.

127 Charles Robin to Philip Robin, 16 November 1797.

128 Charles Robin to Philip Robin, 22 July 1800.

129 The letterbooks give a list of market and supply agents at the beginning of each book.

130 See Fischer and Sager, *The Enterprising Canadians*, especially the papers by Felt, Panting, and Finlay. See also Chapter 3 of this book. The need for detailed research on markets has already been indicated. It would, of course, also help to throw light on kin links.

131 Charles Robin to Philip Robin, 16 November 1797.

132 Charles Robin to Philip Robin, 1 October 1800. John had been in the Lisbon house of Axtell and Robin – the first Robin to be trained in the markets *in situ*.

133 Charles Robin to Burns and Woolsey (Quebec), 10 July 1802. James Robin wrote Burns and Woolsey from Paspébiac on 11 September 1802, "Our Mr. Charles Robin left us 20th past... We sincerely hope he will be favoured with an agreeable passage and happy sight of all our Relations and Friends."

134 See the first two letterbooks.

135 "Nouvelles de mer" in the *Chronique de Jersey*, 1840, 1850; and the Jersey Ship Registers, 1803–50; both sources taken in conjunction.

136 "The Journal of Charles Robin." Raulin Robin was the president of the Jersey Chamber of Commerce in the early 1880s.

137 Where there was no Jersey agent in a market area, firms would deal with a resident local firm, such as Vito, Terni and Company, who had at one time been the Naples agent for the Perrée firm, prior to Dobree, Maingay and Company. Most of the transactions would take place through the main London agent – DeLisle, Janvrin, DeLisle, for example, or Fiott de Gruchy.

138 Chandler, *The Visible Hand*, especially 15–80.

139 Ibid, 15.

CHAPTER 5

1 Quoted in Innis, *The Cod Fisheries*, 287.
2 Lee, *The Robins in Gaspé*; Samson, *Fishermen and Merchants*.
3 Blanchard, *L'est du Canada*; Chambers, *The Fisheries*; Cooney, *A Compendious History*; Langelier, *Esquisse sur la Gaspésie*; Mountain, *Visit to the Gaspé Coast*; Ouellet, *Histoire économique et sociale du Québec*. See also Chaussade, *La Pêche*, and LePage, "Le capitalisme marchand."
4 There are a variety of discussions round this theme; see for example, Gilmour, *The Spatial Evolution*; Vance, *The Merchant's World*; Watkins, "A Staple Theory."
5 Ommer, "What's Wrong with Canadian Fish?," 25–7.
6 See Matthews, "Newfoundland Merchants, Parts I and II"; Ryan, "The Newfoundland Cod Fishery," Appendices, Table 37, 271 ff.; *United Province of Canada*, Sessional Papers, Report on the Canadian Fisheries, 1874, Appendix.
7 These single persons might include the 150 Jerseymen mentioned by the enumerator as being seasonally employed by CRC, and as returning to Jersey each fall.
8 See the "index of permanency" devised by Handcock, "English Migration to Newfoundland," 19–20.

It is worth noting that a comparison of single to married males and females for Gaspé and Bonaventure counties shows that, on average, only 22% of females over the age of fourteen were single, compared to 64% of males in the same age group. Even taking into account the normal later age at marriage of men in this period, these figures suggest that the temporary presence of migratory labour from Jersey inflated the impression of strong demographic growth, as indicated by the *Census* population statistics, to some degree. However, the *Census* exhibits some considerable irregularities in the demographic data: witness the discrepancy between married males and married females in Bonaventure County. The data must therefore be treated with caution, and where possible "household" units have been used instead, calculated from the number of houses given in the *Census*.

9 Watkins, "A Staple Theory," 55.
10 Ibid.
11 See chapter 2.
12 *Report* of J.D. McConnell to F.W. Baddely, *Quebec Mercury*, 18 November 1833. Hereafter cited as "Baddely Report."
13 Fortin Report, 1858. ["The absolute lack of roads has till now pre-

vented the settlers on the coast from establishing inland where the
land is smooth, the soils good and covered in excellent woods."]

14 George LeBoutillier, "Report on the Colonial Roads in Lower
Canada," United Province of Canada, *Sessional Papers*, No. 15, 1861.

15 See Mannion, *Point Lance in Transition*, 27–29 and 36–41, and Thornton, "Dynamic Equilibrium," figure 4.4, 159. Both studies map the
physical layout of the outport.

16 Charles Robin to Philip Robin in Jersey, 22 June 1801, CRC Letterbooks.

17 Charles Robin to Fiott deGruchy in London, 19 December 1795.

18 For linkages derived from shipbuilding, see, for example, Ommer,
"Anticipating the Trend," which looks at development in Pictou, Nova
Scotia. Shipbuilding in Gaspé was useful to Jersey because it circum-
vented the problems of obtaining timber there. Indeed, the transfer of
shipbuilding from the United Kingdom to British North America was
common during the Napoleonic Wars; the classic work remains
Lower, "The Trade in Square Timber." The collapse of shipbuilding by
CRC after the 1830s was partly a result of the removal of these duties.

19 Calculated from the Jersey Ship Registers, 1808–1840.

20 Vance, *The Merchant's World*, 4.

21 Ibid.

22 This is probably true of many systems, and should be typical of, for
example, the forts of the fur trade. The Ohio and Mississippi rivers
may have operated like this at precise points in their historical development.

23 Fortin Report, 1862.

24 Watkins, "A Staple Theory," 55.

25 Baldwin, "Patterns of Development," 161–79; Gilmour, *The Spatial
Evolution*, especially chapter 2; North, "Agriculture in Regional Economic Growth"; Watkins, "A Staple Theory."

26 *Lower Canada, Census and Statistical Returns, 1831*; Gaspé County and
Bonaventure County, General Remarks (20 Will IV, App. 00, 1832).

27 Baddely Report.

28 Ibid.

29 Samson, *Fishermen and Merchants*, 38.

30 No such study exists, and the accounts of visitors to the area tend to
be rather idiosyncratic, and hence unreliable as guides for such an
assessment.

31 Baddely Report.

32 There were six non-Jersey-owned vessels: three whalers and three
small coastal vessels. There were no other ocean-going vessels: see
the Baddely Report.

33 The letterbooks are full of cautions about unnecessary expenditures. See, for example, Jersey to Paspébiac, 20 June 1876, LeGros Papers: "We have a host of horses at Percé. I am suspicious that at Grand River they keep carriage horses, such not wanted and *must* be put a stop to ... Our expenses are heavy enough without all this ... Mr. Orange will see that all this *extravagance* finishes, business cannot support it"; or the following from Raulin Robin to Paspébiac, 16 August 1876, LeGros Papers: "Repairing piano 25/–; who plays on it?"; or C.W. Robin to Paspébiac, 8 November 1865, LeGros Papers: "Mr. Orange must not send his letters in thick envelopes ... we had to pay 1/– extra for the last letter."

34 For a discussion of the technicalities of debt, credit, and the truck system of CRC, see Ommer, "The Truck System," from which parts of this chapter have been drawn.

35 Baddely Report.

36 The exception is 1838, the year LeBoutillier set up in "competition" to CRC, when the number of accounts at Percé fell to just over 150. Rapid recovery resulted in numbers returning to normal by 1842. LeBoutillier's competition was somewhat muted; it was in reality a daughter firm, using CRC's vessels on occasion, and its marketing contacts. Principals of the two firms met regularly in Jersey at the Chamber of Commerce; perhaps "friendly rivalry" would be a more accurate description of their relationship.

37 With the exception of 1838, the year LeBoutillier set up.

38 See Forster, *A Conjunction of Interests*, especially 30–1. In 1855, the Paspébiac agent warned James Robin in Jersey that the Reciprocity Treaty was creating "very critical" conditions on the coast by offering local fishermen cash "on the barrel-head": letter of 5 September 1855.

39 Samson, *Fishermen and Merchants*, 88, shows the same picture for a sample of twenty-five dealers.

40 Robin, Jones and Whitman Papers, Cash Accounts for 1833–63, vols. 175–88.

41 Robin, Jones and Whitman Papers, vols. 172–93, *passim*. The range of premiums charged to clients has been estimated for each year from 1825 to 1877. McConnell underestimated them at 25%, in an *apologia* for the practice (see earlier); Samson cites an example of 35% (*Fishermen and Merchants*, 71); Hilton (*The Truck System*, 40) notes that 10% to 35% was not uncommon in Britain. The evidence is that CRC's premiums were excessive: see Ommer, "The Truck System," for an extended discussion of this point.

42 See Armstrong and Jones, *Business Documents*, 124–33.

43 Baddely Report.

44 Robin, Jones and Whitman Papers, Accounts of: John Peter Arsenault, vols. 173, 175, 178, 181; Charles Doiron, vols. 173, 175, 178; Eduard Huard, vols. 173, 175, 178, 181; Francis Hébert, vols. 173, 175, 178, 181; John Lantin, vols. 173, 175, 178; James Huard, Jr, vols. 173, 175, 178, 181; John Chapados (Michel), vols. 173, 175, 178, 181, 183; Peter Arosbile, vols. 173, 175; Clement Holme, vols. 173, 175, 178, 181, 183; John Sire, vols. 173, 175; Peter Hautseinnet, vols. 173, 175, 178; Andrew Castillon, vols. 173, 175, 178, 181; Rhéné Dugay, vols. 173, 175, 178, 182; Frederick Arsenault, vols. 181, 183, 186, 189–92; Peter E. Arsenault, vols. 183, 186, 189–92; Joseph Arsenault, vols. 181, 183, 186, 189–92; J-B Arsenault, vols. 183, 186, 189–92; Zachary Arsenault, vols. 183, 186, 189–92; Sebastion Arsenault, vols. 181, 183, 186, 189–90; Felix Arsenault, vols. 181, 183, 186, 189–92.

45 Philip Robin was the nephew of Charles Robin, who wed Martha Arbou in Gaspé in a marriage that was concealed (probably for religious reasons) and has never been acknowledged by the family in Jersey: see the marriage and birth certificates in the Gaspé Diocesan Archives. I am grateful to F.W. Remiggi for this information.

46 Robin, Jones and Whitman Papers, Account of James Day, vols. 173, 175.

47 Since CRC controlled the supply trade into the Gaspé fishery, it was at the front of a chain of credit reaching back through middlemen (smaller merchants, such as Hyman – see Samson, *Fishermen and Merchants*, 20–4, – who used CRC to import supplies and export their fish) to the individual on the coast. It was also the firm with the largest store-goods inventory and the greatest financial resources. A debt which CRC was unwilling to carry was clearly even more hazardous for other merchants under these circumstances.

48 Jersey to Paspébiac, 20 June 1876, LeGros Papers.

49 See Ommer, "Merchant Credit and Household Production," for the symbiotic relationship between credit and subsistence in the Newfoundland inshore fishery in the early twentieth century.

50 Chevrier and Blanchard, *Les Défricheurs d'eau*, especially 25–27; Samson, *Fishermen and Merchants*.

51 Cooney, *A Compendious History*.

52 Ibid.

53 Based on export timber and subsequent agricultural development, Ontario could boast, by 1851, that nowhere along the lakeshore was there less than 15% of the total employment of each county working in the secondary sector; by county, 40% or more was employed in the manufacture of producer goods, while in consumer goods it was everywhere over 50%. See Gilmour, *The Spatial Evolution*, 51 and 63. This is not to deny different structural elements in Ontario's growth,

Ontario's growth, but simply to provide a framework of reference and thus to demonstrate the extreme slowness of growth in Gaspé.

54 Overextension of these supply lines led to the collapse of fur companies such as the North-West Company. See Innis, *The Fur Trade in Canada*.

55 Wynn, *Timber Colony*.

56 Fortin Report, 1858; see also Innis, *The Cod Fisheries*, 278–9. ["Nothing lovelier than the order, cleanliness, and efficiency that reigns in these establishments. There is required of the different clerks employed in the fish trade a regular apprenticeship lasting several years. Every senior agent has had charge, over a long time, of a small station where he has had to prove his capacity to work hard; every chief clerk has learned, in more menial positions, to make accurate judgments about the value of goods, the quality of fish."]

57 Baldwin, "Patterns of Development," commented that the "price of the export commodity" (in this case, fish) and "the array of factor prices" (in the mother country) would greatly influence the production function of the staple, which would in turn "greatly affect subsequent development by initially influencing the nature of the labour and capital supply which flows into each region and the distribution of each economy's national income." He demonstrated this by projecting the differences in development between a plantation economy and a wheat economy. In the case of the cod fishery, the export commodity was of low value, the capital requirements relatively low (although beyond the means of local fishermen), and labour cheap.

58 Papineau, for example, is noted by Innis (*The Cod Fisheries*, 281) as commenting that the fishery was "a species of industry the least proper for his country ... They had done quite as much as they needed to in not doing anything to injure the fishermen." His attitude was typical of the French-Canadian élite of the time, who were not interested in the fishery, and (at that point) not even interested in creating alternatives in Gaspé.

CHAPTER 6

1 *The Guernsey and Jersey Magazine*, 306.

2 Ibid.

3 Petition of 1839 to the Board of Trade, Jersey Chamber of Commerce, Minute Books.

4 Minute Books, 18 January 1841.

5 Ibid.

6 Letter from Jersey to the Principal Officers, H.M. Customs, 2 July 1841, "Jersey Letters Inwards": Custom House, London, 156.

7 Minute Books, 1 July 1845, Book v, 14.

8 Ibid.

9 See also April, 1851, Correspondence, Minute Books.

10 Petition to the Lords Commissioners of the Admiralty, Minute Books, 15 March 1854, 125. The Chamber was finally successful in 1860.

11 Minute Books, 7 September 1858, 153.

12 Representation to the Earl of Clarendon (Principal Secretary of State for Foreign Affairs), 14 April 1853. Minute Books, 110.

13 Calculated from *Registered Tonnage of the British Empire*: Mercantile Navy List, 1864, ii, tables 1 and 2.

14 Letter to Lister, 1860, Minute Books, 183.

15 Letter to the Postmaster General, 12 June 1860, Minute Books, 161–3.

16 The percent increase for absolute figures is dramatic, but even less reliable, given the immense discrepancy in the population sizes involved. The figures are: England and Wales 49%, Scotland 38.1%, Guernsey 61%, Isle of Man 30%, and Jersey 100%.

17 The Chamber used this threat of emigration as a weapon with which to assail Westminster whenever it felt the need for English political support. See, for example, Letter of Francis Gifford, 23 March 1816, Minute Books.

18 *The Guernsey and Jersey Magazine*, 306.

19 Ibid., 308.

20 Ibid., 309.

21 Ibid., 370. In the early years, the registers recorded a large preponderance of "prizes" and vessels purchased in places such as the south of England, suggesting that shipbuilding was not then important in Jersey, as figure 19 shows.

22 CRC (Paspébiac) to James MacMillan, Campbelltown, 11 February 1850. CRC Letterbooks.

23 Minute Books, Book V, 1 July 1845, 14.

24 For example, the Mediterranean commodity group clearly served a number of trades, the cod trade and the wine trade being two of these. In this case, the cod trade was essentially market oriented, the wine trade a supply trade to Jersey. See also chapter 4.

25 For the details of this methodology, see Ommer, "Nouvelles de Mer," 147–82.

26 Ragatz, *The Fall of the Planter Class*.

27 See Head, *Eighteenth Century Newfoundland*, 249, where the comment is made that very little is yet known about the trading geometries of the North Atlantic. The comment remains true.

28 More than twenty-five ships were found in the Jersey ship registers from 1803 to 1830 which were known, from sources such as the CRC Letterbooks, to be cod ships, but which were not mentioned in the

Chronique de Jersey. However, many of these were of less than thirty tons: shallops were used as collection boats in the cod fishery. Of the twenty-five counted, nineteen were Newfoundland-built, two were Jersey-built, and there were four "prizes." It appears that under-registration in the cod trade was greater than in other trades; this is not surprising, since the British North American fisheries were the only other area in which Jerseymen owned land in sufficient amounts to develop shipbuilding.

29 In 1830 this is nearly every owner (nineteen out of twenty), and in 1840 still a majority (twenty-six out of thirty-one).

30 Or at least they do not appear in the ship registers, and have therefore been considered here as non-Jersey, although this is perhaps an understatement of Jersey involvement in ship ownership in the Baltic trades. They could have been registered to Jersey owners in another port, such as London, Bristol, Liverpool, or Southampton.

31 This last in partnership with LeBas.

32 The results of this diversification are reflected in the advertisements in local newspapers, inserted by merchants involved in these trades, and offering salt, figs, wine, coffee, sugar, and fisheries produce, to name only a few.

33 Baddely Report.

34 Ibid.

35 *Chronique de Jersey*, 13 March 1830, 21 August 1830, and 11 September 1830.

36 Ibid., 14 August 1830.

37 Ibid., 16 January 1830.

38 Podger, "Shipbuilding in Jersey," 234.

39 "Nouvelles de Mer" of the *Chronique de Jersey* for the years 1840, 1850, and 1860. Also crew lists for Jersey vessels, Maritime History Archive, Memorial University of Newfoundland.

40 Jersey Crew Lists – the *Hasty*, official number 21153, October 1864.

41 Jersey Crew Lists – the *William*, official number 11372, November 1872. This ship was 210 tons, the master was P.J. Marrett, and the managing owner was William Bisson, of St Peter's, Jersey.

42 Jersey Crew Lists.

43 "Nouvelles de Mer" of the *Chronique de Jersey*, and Jersey Ship Registers, for the years 1835 to 1870.

44 *The Guernsey and Jersey Magazine*, 310.

45 Ibid. Also, an interview conducted on 4 November 1977, in St Helier, with J. Norman of J. Norman Ltd., coopers and box-makers.

46 *The Guernsey and Jersey Magazine*, 310–11.

47 Ibid., 313.

48 Ibid., 309.

49 Maritime History Archive.

50 Stevens, *Old Jersey Houses*.

51 Beamish, Hillier, and Johnstone, *Mansions and Merchants*, Vol. 1.

52 Documents in the possession of Guy Janvrin Robin, "Petit Menage," Jersey. Hereafter cited as "Petit Menage Documents." These include a box of Raulin Robin's documents dating to the period of the 1886 bank crash. I am indebted to Guy Janvrin Robin for permission to examine these papers.

53 Stevens, *Victorian Voices*, 235.

54 *The Guernsey and Jersey Magazine*, 312.

55 Petit Menage Documents. See also chapter 3, estate of John Fiott, for capital accumulation in the cod fishery.

56 Baddely Report.

57 Hon. R. B. Sullivan, quoted in Pentland, "The Role of Capital," 459.

58 Interviews: November–December 1976, October–November 1977, with Philip de Veulle, Henry Perrée, Nicholas Robin, Guy Janvrin Robin, and Lady Angela Walker. Interviews conducted in Jersey at the Société Jersiaise or in the homes of the persons listed above, to whom I am greatly indebted.

59 Syvret, "Valpy dit Janvrin." I am indebted to Ms Syvret for giving me a copy of her typescript.

60 Advertisement of sale. Document in the museum of the Société Jersiaise, St Helier.

61 Name Files, Maritime History Archive.

62 LeRossignol, *Notes on Banking*.

63 Perrée Papers.

64 Interview, December 1976, with Philip de Veulle, at Gorey, Jersey.

65 See the publications of the Atlantic Canada Shipping Project at Memorial University, for example, Alexander and Ommer, *Volumes Not Values*; Fischer and Sager, *The Enterprising Canadians*.

CHAPTER 7

1 Williams, *From Columbus to Castro*, 383.

2 Government of Canada, Sessional Papers, *Report on the Canadian Fisheries*, 1886. ["The fishing privileges ceded to the Americans, and the omission of similar privileges to Canadians ... the terms of the Treaty of 1854 with respect to fisheries matters was overly favourable to the United States."]

3 Minute Books of the Jersey Chamber of Commerce.

4 Remiggi, "Nineteenth Century Settlement," 15–43 and especially 23–6; see also 116–18.

5 Government of Canada, *Report on the Canadian Fisheries*, 1867, 38.

6 Paspébiac agent to James Robin (Jersey), 24 June 1850: "The seigneurie is now very thickly settled ... The population is now too great to live exclusively by the fishery." CRC Letterbooks.

7 *Boston Commercial Bulletin*, 28 July 1860.

8 Fortin Report, 1864.

9 Letter of George LeBoutillier, Province of Canada, *Sessional Papers*, No. 15, Appendix No. 35, 1861.

10 Remiggi, "Nineteenth Century Settlement," 15–43, especially 27–43; see also 271–7.

11 Government of Canada, *Report on the Canadian Fisheries*, 1865, 38. ["It was a new era for this country: now it would not take long to make rapid progress."]

12 Syvret, "Valpy dit Janvrin."

13 Government of Canada, *Report on the Canadian Fisheries*, 1865. ["It is not now (and can no longer be) the same way of doing business as before... This year, to my certain knowledge, there have been many sales of fish for thousands of louis, in hard cash."]

14 Ommer, "The decline," which deals with the international context of the shift from sail to steam over the second half of the nineteenth century.

15 *Mitchell's Maritime Register*, 16 July 1870.

16 Ibid., 2 April 1870.

17 Ibid., 5 February 1870; Bombay report of 8 January 1870.

18 Ibid., 25 June 1870.

19 Minute Books, Annual General Meeting (AGM), 1867. This collapse of sail, and of freight rates, was not felt only in Jersey: see Ommer, "The decline."

20 See Landès, *The Unbound Prometheus*.

21 *Mitchell's Maritime Register*, 28 July 1860.

22 Williams, *From Columbus to Castro*, 383.

23 Ibid.

24 Minute Books, 1865–73, *passim*.

25 Ibid.

26 Minute Books, AGM, 1873.

27 Minute Books, AGM, 1879.

28 Minute Books, AGM, 1880.

29 Ibid.

30 Ibid.

31 Minute Books, AGM, 1881.

32 Minute Books, AGM, 1884.

33 Minute Books, AGM, 1885.

34 *Nouvelle Chronique de Jersey*, 11 January 1886.
35 Petit Ménage Documents. Interview with Guy Janvrin Robin, October 1977, at Petit Ménage. This means "bankrupt."
36 *Nouvelle Chronique de Jersey*, 11 January 1886.
37 Minute Books, AGM, 1887.
38 Minute Books, 1887 onward.
39 *Nouvelle Chronique de Jersey*, November 1886; a series of advertisements throughout the month.
40 *Nouvelle Chronique de Jersey*, 11 January 1886. Contents of the London report on the bank crash. ["In a financial upheaval of the kind that we are appalled by today ... it remains to us ... to counsel ... prudence and patience ... One respects a general who has lost a battle. One deplores his fall, sheds a sympathetic tear, knowing that he engaged in a good cause. By analogy, then, one must surround the name of Robin with the greatest possible respect and the deepest sympathy." "Our generation has witnessed great events. Wooden ships have been replaced by iron. Sail has been replaced by steam. The telegraph has taken the place of correspondence. The old order is passing, and new things are coming to pass."]
41 Sutherland, "The Personnel and Policies"; see also McCann, "Staples and the New Industrialism."
42 Letter from Paspébiac agent to Robert Christie (Toronto), MPP for Gaspé, over a dispute with the Crown Lands Office, 13 July 1850.
43 Innis, *The Cod Fisheries*, 507.
44 Ibid., 508.

CHAPTER 8

1 Saxe, "The Blind Men and the Elephant."
2 See, for example, Antler, "Colonial Exploitation"; Baldwin, "Patterns of Development"; Birnberg and Resnick, *Colonial Development*; Buckley, "The Role of the Staple Industries"; Caves and Holton, *The Canadian Economy*; Denoon, *Settler Capitalism*; Drache, "Rediscovering Canadian Political Economy"; Gilmour, *The Spatial Evolution*; Innis, *The Cod Fisheries*; Naylor, "The History of Domestic and Foreign Capital in Canada"; North, *The Economic Growth*; Vance, *The Merchant's World*; Watkins, "A Staple Theory."
3 See, for example, the essays in Easterbrook and Watkins, *Approaches to Canadian Economic History*.
4 Baldwin, "Patterns of Development."
5 Ibid.; also North, *The Economic Growth*.
6 Gilmour, *The Spatial Evolution*, and Watkins, "A Staple Theory."

7 Some good examples of sensitivity to this wider context can be found in the essays by Bosquet, Cronin, Bergesen, Clay, and Ragin in Hopkins and Wallerstein (eds.), *Processes of the World-System*, and those by Montejano, Lubeck, and Milkman in Goldfrank (ed.), *The World-System of Capitalism*. See also Ommer, "'The Misery of Our Poor'".

8 Paquet, "Some Views," 44.

9 See Drache, "Rediscovering Canadian Political Economy"; Naylor, "The History of Domestic and Foreign Capital"; and the extensive literature on development and underdevelopment, including such classics as Frank, "The Development of Underdevelopment"; Furtado, *Economic Development of Latin America*; Griffen, *Underdevelopment in Spanish America*; and Sideri, *Trade and Power*. See also Fox-Genovese and Genovese, *Fruits of Merchant Capital*.

10 Easterbrook, "Problems in the Relationship of Communication and Economic History," 563.

11 Gilmour, *The Spatial Evolution*, 20.

12 Price, "The Transatlantic Economy," 24.

Bibliography

MAJOR PRIMARY SOURCES:
DOCUMENTS, GOVERNMENT
REPORTS AND PUBLICATIONS,
CENSUSES, NEWSPAPERS

Boston Commercial Bulletin, 28 July 1860. Microfilm, in the Maritime History Archive, Memorial University of Newfoundland, St John's, Newfoundland.

Robin, Jones and Whitman Papers. There is an extensive collection of documents belonging to CRC in the National Archives, Ottawa, under MG 28 III 18. Used in this study are the Letterbooks (microfilmed: reels M903 and M904) and the Ledgers, vols. 172–93. There is also a smaller collection of CRC papers, mostly copies of the National Archives material, in the Public Archives of Nova Scotia. Beyond this, there is some material held privately: see LeGros Papers and Petit Menage Documents, this bibliography.

Chronique de Jersey, 1820 to 1887 (later called the *Nouvelle Chronique de Jersey*). Collection in the Société Jersiaise, St Helier, Jersey.

Crew Lists of Jersey Shipping. Lodged in the Maritime History Archive, Memorial University of Newfoundland, St John's, Newfoundland.

DeCarteret and LeVescomte Papers, Volume 1, 1829–89. MG 1, 257. Public Archives of Nova Scotia.

Fiott Papers. A collection of papers, principally around 1770, of the Fiott family of Jersey. Lodged in the Société Jersiaise, St Helier, Jersey.

The Guernsey and Jersey Magazine. Jersey 1837. Held in the Jersey Library, St Helier, Jersey.

Glenalladale Papers. A Collection of papers of John MacDonald of Glenalladale. Provincial Archives of Prince Edward Island, Charlottetown.

Haldimand Papers. Haldimand Collection MG 21 G 2 "B" Series, Vol. 202 (1774–84). National Archives, Ottawa.

Harbour Grace Clearances, 1776–94, GN 11. Provincial Archives of New-foundland.

Jersey Chamber of Commerce Minute Books. Chamber of Commerce, St Helier, Jersey.

Jersey Letters Inwards. Letterbooks of Her Majesty's Customs, 1841–87. Lodged at the Custom House, London, England.

Jersey Ship Registers, 1803–87. Originals held in the Bureau des Impôts, Weighbridge, St Helier, Jersey.

Journal of Charles Robin. Original in the Société Jersiaise, St Helier, Jersey. Copy in the National Archives, Ottawa: MG 23 G 111 24.

Journal de Daniel Messervy, 1769–72. Jersey: Société Jersiaise 1896.

Journals of the House of Assembly of Newfoundland, 1859. St John's, Newfoundland.

LeGros Papers. Documents, typescript, and photostat copies of originals (now in the National Archives), along with boxes of letters of the firm of CRC. In the possession of the family of the late Arthur LeGros of Paspébiac, Gaspé, Quebec.

Lower Canada, Census and Statistical Returns. For Gaspé County and Bonaventure Country, 1831 and 1851. National Archives, Ottawa.

Mitchell's Maritime Register, 1860–75. Microfilm in the archives of the Maritime History Group, Memorial University of Newfoundland, St John's, Newfoundland.

Name Files. A collection of information on surnames (individuals and families) of Newfoundland. Compiled by Keith Matthews, and lodged in the Maritime History Archive, Memorial University of Newfoundland, St John's, Newfoundland.

Navigating Troubled Waters: Report of the Kirby Task Force on the Atlantic Fishery. Ottawa: Queen's Printer 1983.

Perrée Papers. Papers held in part by Jurat Henry Perrée of Jersey, and partly by the Société Jersiaise, St Helier, Jersey. The documents consist of a series of letters written circa 1850, and a ship account book for the same period.

Petit Menage Documents: In the possession of Guy Janvrin Robin, of Petit Menage, Jersey. The collection is mostly papers, dated circa 1886, containing the correspondence of Raulin Robin of CRC about events surrounding the 1886 bank crash.

Report on the Canadian Archives, 1888: Report on the Haldimand Collection. Ottawa: Queen's Printer 1889.

Report of J.D. McConnell to F.W. Baddely. *Quebec Mercury,* 18 November 1833. National Archives, Ottawa.

The Shipping and Mercantile Gazette. London 1855. On microfilm in the Maritime History Archive, Memorial University of Newfoundland, St John's, Newfoundland.

United Province of Canada/Government of Canada, Sessional Papers (Documents de la Session). "Report on the Canadian fisheries in the Gulf of St Lawrence." Appendices. 1854–87. National Archives, Ottawa.

United Province of Canada, Sessional Papers, "Procédés sous l'Acte Seigneurial de 1854 et ses amendements." A series of cadastral listings of seigneuries in the Gaspé area, and of the residents there. 1858. National Archives, Ottawa.

United Province of Canada, Sessional Papers. "Report on the Colonial Roads in Lower Canada." No. 15, 1861. National Archives, Ottawa.

SECONDARY SOURCES:
BOOKS AND ARTICLES

Acheson, T.W. "The Great Merchant and Economic Development in Saint John, 1820–1850." In Bercuson, D.J., and Buckner, P.A., eds., *Eastern and Western Perspectives*, 85–114. Toronto: University of Toronto Press 1981.

Alexander, David. *The Decay of Trade*. St John's: Institute of Social and Economic Research, No. 19, Memorial University of Newfoundland 1977.

– "Newfoundland's Traditional Economy and Development to 1934." In Hiller, J.K., and Neary, P., eds., *Newfoundland in the Nineteenth and Twentieth Centuries*, 17–39. Toronto: University of Toronto Press 1980.

– "Economic Growth in the Atlantic Region, 1880–1940." In Bercuson, D.J., and Buckner, P.A., eds., *Eastern and Western Perspectives*, 197–227. Toronto: University of Toronto Press 1981.

Alexander, D., and Ommer, R. *Volumes Not Values*. St John's: Maritime History Group, Memorial University of Newfoundland 1980.

Andrews, Charles M. *The Colonial Period of American History*. 3 vols. New Haven: Yale University Press 1934–38.

Antler, Steven David. "Colonial Exploitation and Economic Stagnation in 19th Century Newfoundland." Ph.D. diss., University of Connecticut 1975.

Bailyn, Bernard. *The New England Merchants in the Seventeenth Century*. Boston: Harvard University Press 1955.

Balcom, D.A. "Production and Marketing in Nova Scotia's Dried Fish Trade, 1850–1914." MA thesis, Memorial University of Newfoundland 1981.

Baldwin, R.E. "Patterns of Development in Newly Settled Regions." *Manchester School of Economic and Social Studies* 24 (1956): 161–79.

Balleine, G.R. *A History of the Island of Jersey*. London: Hodder and Stoughton 1950.

Beamish, D., Hillier, J., and Johnstone, H.F.V. *Mansions and Merchants of Poole and Dorset*. 2 vols. Poole: Poole Historical Trust 1976.

Bertram, G.W. "Economic Growth in Canadian Industry, 1870–1915: The

Staple Model and the Take off Hypothesis." *Canadian Journal of Economics and Political Science* 29 (1963): 159–84.

Birnberg, T.B., and Resnick, S.A. *Colonial Development: An Econometric Study.* New Haven: Yale University Press 1975.

Blanchard, Raoul. *L'est du Canada français.* In three parts. Montreal: Librairie Beauchemin 1935.

Bloch, Marc. *"Les rois thaumaturges": Étude sur le caractère surnaturel attribué à la puissance royale, particulièrement en France et en Angleterre,* Paris: A. Colin 1961.

Brière, J.F. "Le trafic terre-neuvier malouin dans la première moitié du XVIIIe siècle 1713–1755." *Histoire Sociale/Social History* 2 (1978): 356–74.

Brookfield, Harold. *Interdependent Development.* London: Methuen 1975.

– *Colonialism, Development and Independence: The case of the Melanesian Islands in the South Pacific.* Cambridge: Cambridge University Press 1972.

Buckley, K. "The Role of the Staple Industries in Canada's Economic Development." *Journal of Economic History* 18 (1958): 439–50.

Caves, R.E., and Holton, R.H. *The Canadian Economy: Prospect and Retrospect.* Cambridge, Mass.: Harvard University Press 1961.

Cell, Gillian. *English Enterprise in Newfoundland, 1577–1660.* Toronto: University of Toronto Press 1969.

Chambers, E.T.D. *The Fisheries of the Province of Quebec.* Quebec: Department of Colonization, Mines and Fisheries of the Province of Quebec 1912.

Chandler, Jr., A.D. *The Visible Hand: the Managerial Revolution in American Business.* Cambridge, Mass.: Belknap Press 1977.

Chang, M. "Newfoundland in Transition: the Newfoundland Trade and Robert Newman and Company, 1780–1805." MA thesis, Memorial University of Newfoundland 1975.

Chaussade, Jean. *La Pêche et les Pêcheurs des Provinces Maritimes du Canada.* Quebec City: Les presses de l'Université de Montréal 1983.

Chevrier, C., and Blanchard, L. *Les Défricheurs d'eau.* Caraquet: Village Historique Acadien 1978.

Cook, Michael. *The Headguts and Soundbone Dance.* St John's: Breakwater Books 1973.

Cooney, Robert. *A Compendious History of the Northern Part of the Province of New Brunswick and of the District of Gaspé in Lower Canada.* Chatham, Miramichi: D.G. Smith 1896.

DeGruchy, G.F.B. *Medieval Land Tenures in Jersey.* Jersey: Don Balleine Trust 1957.

De la Morandière, Charles. *Histoire de la pêche française de la morue dans l'Amérique septentrionale.* 3 vols. Paris: Sorbonne 1962–63.

Denoon, Donald. *Settler Capitalism: The Dynamics of Dependent Development in the Southern Hemisphere.* Oxford: Oxford University Press 1983.

Drache, Daniel. "Rediscovering Canadian Political Economy." *Journal of Canadian Studies* 11 (August 1976): 13–18.

Easterbrook, W.T. "Problems in the Relationship of Communication and Economic History." *Journal of Economic History* 20 (1960): 559–65.

Easterbrook, W.T., and Aitken, H.G.J. *Canadian Economic History*. Toronto: University of Toronto Press 1956.

Easterbrook, W.T., and Watkins, M.H., eds., *Approaches to Canadian Economic History*. Toronto: McClelland and Stewart 1967.

Egnal, M., and Ernst, J.A. "An Economic Interpretation of the American Revolution." *William and Mary Quarterly*, 3rd series, 29 (1972).

Falle, Philip. *An Account of the Island of Jersey*. Jersey 1837.

Faris, James C. *Cat Harbour: A Newfoundland Fishing Settlement*. St John's: Institute of Social and Economic Research, No. 3, Memorial University of Newfoundland 1972.

Fay, C.R. "South American and Imperial Problems." *University of Toronto Quarterly* (January 1932): 183–96.

Ferland, J. *Cours d'histoire du Canada*. 2 vols. Toronto: Clarke, Irwin & Co. 1969.

Fischer, L.R., and Sager, E.W., eds., *The Enterprising Canadians: Entrepreneurs and Economic Development in Eastern Canada, 1820–1914*. St John's: Maritime History Group, Memorial University of Newfoundland 1979.

Forster, Benjamin. *A Conjunction of Interests: Business, Politics, and Tariffs 1825–1879*. Toronto: University of Toronto Press 1986.

Fox-Genovese, E., and Genovese, E.D. *Fruits of Merchant Capitalism*. Oxford: Oxford University Press 1983.

Frank, André Gunder. "The Development of Underdevelopment." *Monthly Review* 18, no. 4 (1966).

– *Capitalism and Underdevelopment in Latin America*. New York: Monthly Review Press 1969.

Frank, David. "The Cape Breton Coal Industry and the Rise and Fall of the British Empire Steel Corporation." *Acadiensis* 7, no. 1 (1977): 3–34.

Furtado, C. *Economic Development of Latin America: A Survey of Colonial Times to the Cuban Revolution*. Cambridge: Cambridge University Press 1970.

Gilmour, James M. *Spatial Evolution of Manufacturing, Southern Ontario, 1851–1891*. Toronto: University of Toronto Press 1972.

Girvan, N. "The Development of Dependency Economics in the Caribbean and Latin America: Review and Comparison." *Social and Economic Studies* 22 (1973): 1–33.

Goldfrank, W.L. (ed.). *The World-System of Capitalism: Past and Present*. Beverley Hills: Sage Publications 1979.

Gordon, H. Scott. "The Economic Theory of a Common-Property Resource: the Fishery." *Journal of Political Economy* 62 (1954): 124–42.

Griffin, K. *Underdevelopment in Spanish America*. London: Allen and Unwin 1969.

Head, C. Grant. *Eighteenth Century Newfoundland: A Geographer's Perspective*. Toronto: McClelland and Stewart, Carleton Library, No. 99, 1976.

Heyting, W.J. *The Constitutional Relationship Between Jersey and the United Kingdom*. Jersey: The Jersey Constitutional Association 1977.

Hilton, George W. *The Truck System, Including a History of the British Truck Acts, 1465–1960*. Westport, Conn.: Greenwood Press 1960.

Hopkins, T.K., and Wallerstein, I., eds., *Processes of the World System*. Beverley Hills: Sage Publications 1980.

Hubert, Paul. *Les Îles de la Madeleine et les Madelinots*. Rimouski 1926.

Imlah, A.H. *Economic Elements in the Pax Britannica*. Cambridge, Mass.: Harvard University Press 1958.

Innis, Harold A. *The Fur Trade in Canada: An Introduction to Canadian Economic History*. Toronto: University of Toronto Press, revised edition 1956.

– *The Cod Fisheries: the History of an International Economy*. Toronto: University of Toronto Press, revised edition, 1954.

Jean, Yves. "Seasonal Distribution of Cod (*Gadus morhua* L.) along the Canadian Atlantic Coast in Relation to Water Temperature." *Journal of the Fisheries Research Board of Canada* 21, no. 3 (1964): 429–60.

Landès, D.S. *The Unbound Prometheus: Technological Change and Industrial Development in Western Europe from 1750 to the Present*. Cambridge: Cambridge University Press 1969.

Langelier, J.C. *Esquisse sur la Gaspésie*. Quebec 1884.

Lee, David. *The Robins in Gaspé. 1766–1825*. Ontario: Fitzhenry and White-side 1984.

Lefeuvre, George F. *Jèrri Jadis*. Jersey: Don Balleine Trust 1973.

– *Histouaithes et Gens d'Jèrri*. Jersey: Don Balleine Trust 1976.

LePage, André. "Le capitalisme marchand et la pêche à la morue en Gaspésie; la Charles Robin and Company dans la baie des Chaleurs (1820–1870)." Ph.D. diss., Université Laval 1983.

Lequesne, Charles. *Constitutional History of Jersey*. Jersey 1856.

LeRossignol, S.J. *Notes on Banking and Political Events in Jersey*. Trowbridge: B. Landowne and Sons 1915.

Mannion, J.J., ed. *The Peopling of Newfoundland*. Toronto: University of Toronto Press 1977.

– *Point Lance in Transition*. Toronto: McClelland and Stewart 1976.

Marx, K., and Engels, F. *The Communist Manifesto*. London: Penguin 1967.

Matthews, Keith. "A History of the West of England–Newfoundland Fishery." Ph.D. diss., University of Oxford 1968.

– *Lectures on the History of Newfoundland 1500–1830*. St John's: Breakwater Books 1988.

- "Pipon Family." Monograph. St John's: Maritime History Group, Memorial University of Newfoundland 1974.

McCallum, John. *Unequal Beginnings: Agriculture and Economic Development in Quebec and Ontario until 1870.* Toronto: University of Toronto Press 1980.

McCann, Larry. "Staples and the New Industrialism in the Growth of Post-Confederation Halifax." *Acadiensis* 8 (1979): 47–79.

McCusker, John J., and Menard, Russell R. *The Economy of British America, 1607–1789.* Chapel Hill: University of North Carolina Press 1985.

Mountain, G.J. *Visit to the Gaspé Coast.* Quebec City: Archives de la Province du Québec 1943.

Naylor, T. "The History of Domestic and Foreign Capital in Canada." In Laxer, R., ed., *(Canada) Ltd.,* 42–56. Toronto: McClelland and Stewart 1973.

North, D.C. "Agriculture in Regional Economic Growth." *Journal of Farm Economics* 41, no. 5 (1959): 943–51.

- *The Economic Growth of the United States, 1790–1860.* New York: W.W. Norton and Company 1961.

Ommer, R.E. "Scots Kinship, Migration and Early Settlement in Southwestern Newfoundland." MA thesis, Memorial University of Newfoundland 1974.

- "'All the Fish of the Post': Resource Property Rights and Development in a Nineteenth-Century Inshore Fishery." *Acadiensis* 10, no. 2 (1981): 107–23.

- "'The Misery of Our Poor': recent studies in development literature." *Labour/LeTravailleur* 11 (1983): 196–203.

- "A Peculiar and Immediate Dependency of the Crown." *Business History* 25, no. 2 (1983): 107–24.

- "The decline of the eastern Canadian shipping industry, 1880–95." *Journal of Transport History* Third Series, 5, no. 1 (1984): 25–44.

- "Merchant Credit and Household Production in Newfoundland, 1918–28." Paper presented to the Canadian Historical Association, Quebec City, June 1989.

- "The Truck System in Gaspé, 1822–77." *Acadiensis* 19, no. 1 (1989): 91–114.

Ouellet, Fernand. *Histoire économique et sociale du Québec, 1760–1850, Structures et conjoncture.* Montreal: Fides 1966.

Paquet, G. "Some Views on the Pattern of Economic Development." In T.N. Brewis, ed., *Growth and the Canadian Economy,* 34–64. Carleton Library No. 39. Toronto: McClelland and Stewart 1968.

Pentland, H. Clare. *Labour and Capital in Canada, 1650–1860.* Toronto: James Lorimer and Sons 1981.

– "The Role of Capital in Canadian Economic Development before 1815." *Canadian Journal of Economic and Political Science* 16, no. 4 (1950): 457–74.

Podger, A. "Shipbuilding in Jersey." *Annual Bulletin of the Société Jersiaise* (1975).

Poigndestre, Jean. *Caesarea, or a Discourse on the Island of Jersey Circa 1682.* St Helier: Société Jersiaise 1889.

Pollock, F., and Maitland, F.W. *The History of English Law.* 2 vols. Cambridge: Cambridge University Press 1911.

Price, Jacob. "The Transatlantic Economy." In Greene, J.P., and Pole, J.R., eds., *Colonial British America*, 18–42. Baltimore: The Johns Hopkins Press 1984.

Prowse, D.W. *A History of Newfoundland from the English Colonial and Foreign Records.* London 1896. Reprint. Belleville, ON: Mika Studio 1972.

Quinn, D.B. "Newfoundland in the Consciousness of Europe in the Sixteenth and Early Seventeenth Centuries." In Story, G. M., ed., *Early European Settlement and Exploitation in Atlantic Canada*, 3–30. St John's: Memorial University of Newfoundland 1982.

Ragatz, L. *The Fall of the Planter Class in the British Caribbean, 1763–1863.* 1963. Reprint. New York: The Century Co. 1928.

Remiggi, Frank William. "Nineteenth Century Settlement and Colonization on the Gaspé North Coast: An Historical-Geographical Interpretation." Ph.D. diss., McGill University 1983.

Revue d'Histoire de la Gaspésie 16, nos 2 and 3 (1978). Special issue devoted to the Jersey presence in Gaspé.

Ryan, Shannon. "The Newfoundland Cod Fishery in the Nineteenth Century." MA thesis, Memorial University of Newfoundland 1971.

– *Fish Out of Water. The Newfoundland Saltfish Trade 1814–1914.* St John's: Breakwater Books 1986.

Samson, Roch. *Fishermen and Merchants in 19th Century Gaspé.* Hull, Quebec: Supply and Services Canada 1984.

Saunders, A.C. *Jersey in the 18th and 19th Centuries.* Jersey 1930.

Saxe, John G. "The Blind Men and the Elephant." In Woods, Ralph L., *A Treasury of the Familiar*, 8–9. New York: McMillan 1943.

Sider, G.M. "Christmas Mumming and the New Year in Outport Newfoundland." *Past and Present*, 71 (1976): 102–25.

Sideri, S. *Trade and Power*, Rotterdam: Rotterdam University Press 1970.

Smith, Adam. *An Inquiry into the Nature and Causes of the Wealth of Nations.* Oxford: Oxford University Press 1976.

Stevens, J. *Old Jersey Houses.* Jersey: Société Jersiaise 1963.

– *Victorian Voices.* Jersey: Société Jersiaise 1967.

Story, G.M. (ed.). *Early European Settlement and Exploitation in Atlantic Canada.* St John's: Memorial University of Newfoundland 1982.

Sutherland, David A. "The Personnel and Policies of the Halifax Board of Trade." In Fischer, L.R., and Sager, E.W., eds., *The Enterprising Canadians*. St John's: Maritime History Group, Memorial University of Newfoundland 1979.

Syvret, M. "Valpy dit Janvrin." Manuscript in the possession of the Société Jersiaise, St Helier, Jersey.

Templeman, W. *Bulletin No. 154*. Ottawa: Fisheries Research Board of Canada 1966.

Thornton, Patricia A. "Demographic and Mercantile Bases of Initial Permanent Settlement in the Strait of Belle Isle." In Mannion, J. J., ed., *The Peopling of Newfoundland*, 152–83. Toronto: University of Toronto Press 1977.

– "Dynamic Equilibrium: Settlement, Population and Ecology in the Strait of Belle Isle, Newfoundland, 1840–1940." Ph.D. diss., University of Aberdeen 1978.

Turgeon, Laurier. "Pour une histoire de la pêche: le marché de la morue à Marseille au xviiie siècle." *Histoire Sociale/Social History* 14 (1981): 295–322.

Vance, Jr., James E. *The Merchant's World: The Geography of Wholesaling*. Englewood Cliffs, New Jersey: Prentice Hall 1970.

Wade, M. "The Loyalists and the Acadians." *Proceedings of the French in New England, Acadia and Québec Conference*, Orono, Maine (1972): 7–9.

Wallerstein, I. *Political Economy of World-System Annuals*, Beverley Hills: Sage Publications 1978, *passim*.

Watkins, M. "A Staple Theory of Economic Growth." *Canadian Journal of Economics and Political Science* 29, no. 2 (1963): 141–58.

Williams, Eric. *From Columbus to Castro: The History of the Caribbean, 1492–1969*. New York: André Deutsch 1970.

Wright, Louis B. *The Atlantic Frontier: Colonial American Civilization, 1607–1763*. New York: Knopf 1970.

Wynn, Graeme. *Timber Colony*. Toronto: University of Toronto Press 1981.

Index